Bitesize
AQA GCSE (9-1)
GEOGRAPHY
REVISION WORKBOOK

Series Consultant:
Harry Smith

Author:
Michael Chiles

Contents

✓ Tick off each topic as you go.

 Each bite-sized chunk has a **timer** to indicate how long it will take. Use them to plan your revision sessions.

 Scan the **QR codes** to visit the BBC Bitesize website. It will link straight through to revision resources on that subject. You can also access these by visiting www.pearsonschools.co.uk/BBCBitesizeLinks.

Defining natural hazards

② Quick quiz

1. Select **two** climatic hazards from the following:

earthquake ☐ drought ☐

tropical storm ☐ volcanic eruption ☐

2. Complete the sentence using the words below.

A natural hazard is an extreme event that

can have ...., social and

environmental ....

economic   locations   planning   consequences

⑤ Natural hazards   Grade 2

1. Study **Figure 1**, a photo of a natural hazard.
Suggest the hazard shown in this photograph.   **[2 marks]**

....

2. Complete the following sentence.   **[2 marks]**

Hazard risk is increasing because the proportion of the

global ....population...... living in seismically active areas

is ....

Figure 1 Tectonic and climatic hazards

⑩ Factors affecting natural hazard risks   Grade 4

3. Which **one** of the following statements describes the effect of climate change on natural hazard risk?
Shade **one** circle only.   **[1 mark]**

A Climate change is not having any effect on natural hazard risk.   ○

B Climate change is causing a greater risk of tectonic hazards in most areas.   ○

C Rising global temperatures are increasing the frequency and intensity of natural   ○
hazards such as floods, landslides, and tropical storms.

D Climate change is not changing the number of people affected by natural hazards.   ○

4. Explain how the location of a natural hazard can increase the risk to people.   **[2 marks]**

....

....

5. Explain how urbanisation can cause increased natural hazard risk.   **[2 marks]**

....

....

....

....

Exam focus

For a 2-mark question you do not need to include a
lot of detail. Make one briefly developed point.

Distribution of earthquakes and volcanoes

⑤ Quick quiz

1. Which type of crust is most dense? Circle the correct answer.

continental   oceanic

2. List the **four** layers of the Earth, from the centre of the Earth outwards.

....

....

3. Complete the sentences using the words below.

Continental crust is .... and formed from .... rocks.

Oceanic crust is .... and formed from .... rocks.

| thick | thin | younger | older |

⑮ Distribution of tectonic hazards   Grade 4

1. Study **Figure 1**, a map of volcanoes and plate boundaries. Which of the following statements are true? Shade **two** circles only.   **[2 marks]**

A The plate boundary between the South American plate and the Nazca plate is a destructive boundary. ⬭

B Volcanoes are most common around the equator. ⬭

C The plate boundary between the North American plate and the Eurasian plate is a constructive boundary. ⬭

D Most volcanoes in Africa are located on the west coast. ⬭

E The Pacific plate mostly has constructive plate boundaries. ⬭

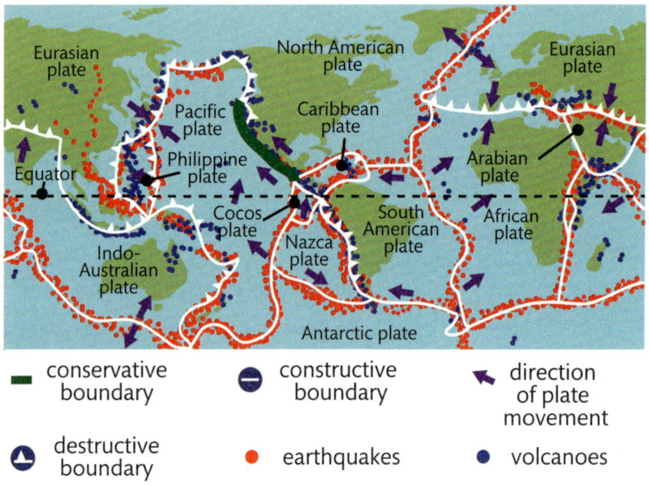

2. Study **Figure 1**. Describe the distribution of volcanoes in South America.   **[2 marks]**

....

....

....

....

Figure 1

Geographical skills 🌐

When describing patterns on a map you need to consider whether the volcanoes are arranged in a scattered, nucleated or linear pattern.

3. Using **Figure 1**, explain how tectonic plate movement causes earthquakes and volcanic eruptions.   **[2 marks]**

🚩 *Convection currents in the Earth's mantle cause tectonic plates to move. This generates* ....

....

4. State what is meant by the term 'intraplate earthquakes'.   **[1 mark]**

Intraplate earthquakes do not occur in the same places as most earthquake events.

🚩 *Intraplate earthquakes are* ....

 Made a start   Feeling confident   Exam ready

Plate margin activity

② Quick quiz

1. Which tectonic hazards do not occur at conservative plate margins?

....

2. At which type of plate margin are volcanic islands formed?

....

⑤ Plate margins
Grade 3

1. Study **Figure 1**, which shows a plate margin. State the type of crust shown at **X** and **Y**. **[2 marks]**

Feature X

oceanic crust ....

Feature Y

....

2. Identify the type of plate margin shown. **[1 mark]**

....

....

3. Study **Figure 1**. State **one** geographical feature that forms at this plate margin. **[1 mark]**

....

Figure 1

⑩ Causes of earthquakes
Grades 5–9

4. Explain how earthquakes occur at destructive plate margins. **[6 marks]**

....
....
....
....
....
....
....
....
....
....
....
....

Exam focus 📌

For a 6-mark question, provide detailed explanations that demonstrate your knowledge and understanding. You will need to use specialist terminology accurately to access the highest marks.

Named example: Earthquake effects

② Quick quiz

1. Complete the sentence by ticking **one** of the options below.

A primary effect is...

☐ ...the first time an earthquake happens in a country.

☐ ...something that happens as a direct result of a tectonic hazard.

2. Is this statement true or false? Circle your answer.

Wealthier countries are normally better prepared for the effects of earthquakes than poor countries.

true   false

⑩ The impact of earthquakes   Grade 5

1. State **one** primary effect of an earthquake.   **[1 mark]**

....

2. State **two** secondary effects of an earthquake.   **[4 marks]**

Effect 1 *Earthquakes can cause people to lose key services such as electricity or phone lines, and can cause mains water pipes to burst.*

Effect 2 ....

....

3. Explain **one** reason why the effects of an earthquake tend to be more severe in a lower income country (LIC) compared to a higher income country (HIC).   **[2 marks]**

....
....
....
....

📍 Named example

Use named examples to show how the effects vary between two areas of contrasting wealth. Revise your named examples at the same time, and learn the same key facts for both, such as the year, the number of people killed and the number of people made homeless. Use these facts to answer comparison questions.

⑤ Primary effects   Grade 5

4. Using an example you have studied, explain how the primary effects of an earthquake can affect people.   **[4 marks]**

....
....

📌 Exam focus

For a 4-mark question, provide two points supported with evidence.

....
....
....

 Made a start   **Feeling confident**   ☐ **Exam ready**

Named example: Earthquake responses

② Quick quiz

1. Which of the following can cause earthquakes?
Tick **two** answers only.

☐ Underground mining

☐ Tsunamis

☐ Volcanoes

☐ Movement of tectonic plates

2. Complete the sentence by ticking **one** of the options below.

The deployment of soldiers for search and rescue after an earthquake is...

☐ ...an immediate response.

☐ ...a long-term response.

⑩ Long-term responses · Grade 5

1. Outline **two** long-term responses to an earthquake hazard in an urban area. **[4 marks]**

Response 1 *One long-term response to an earthquake is the NGO Oxfam working with local communities to raise awareness of good hygiene practices to limit the spread of waterborne diseases.*

Response 2 ....
....

2. Using examples you have studied, explain why the long-term responses to an earthquake vary in **two** areas of contrasting wealth. **[6 marks]**

....
....
....
....
....
....
....
....
....

Named example 📍

When revising the two named examples you have studied, consider why the responses were different in the two countries.

When recovering from a tectonic hazard, HICs can afford to design and construct earthquake-resistant buildings, but poorer countries cannot afford to do this.

⑤ Immediate responses · Grade 5

3. For an example you have studied, outline **two** immediate responses to an earthquake. **[4 marks]**

Response 1 ....
....

Response 2 ....
....

Living with natural hazards

(2) Quick quiz

1. Is this statement true or false? Circle your answer.

The infrequency of hazards means some people believe they will not be affected in their lifetime.

true   false

2. Complete the sentence using the words below.

Volcanic eruptions can sometimes result in .... crop yields due to minerals in the ash.

increased   decreased

(5) Living with tectonic hazards   Grade 4

1. Suggest how volcanoes can provide opportunities for people.   **[2 marks]**

Areas with volcanic activity can provide opportunities to generate ....

....

2. Outline **one** economic benefit of living in hazardous zones.   **[2 marks]**

....

....

(15) Living in hazardous zones   Grade 5–9

3. 'The benefits of living in hazardous zones outweigh the risks.'
To what extent do you agree with this statement?   **[9 marks]**
                                                        [+ 3 SPaG marks]

....

....

....

....

....

....

....

....

....

....

....

....

....

Exam focus

The command phrase **to what extent** means you need to say how far you agree or disagree. To achieve the higher marks in this question, write about the benefits and risks, use specific examples and finish by stating how much you agree or disagree with the statement.

Exam focus

This question has 3 extra marks available for SPaG. Make sure that your answer is well-structured and includes specialist geographical terms. Read it through at the end to check your spelling, punctuation and grammar.

Reducing the risk

② Quick quiz

1. Name **two** tectonic hazard monitoring devices.

....

....

2. Which hazards are more difficult to predict accurately? Circle the correct answer.

earthquakes   volcanic eruptions

⑤ Hazard mapping   Grade 4

1. Study **Figure 1**, which shows a tectonic hazard map of New Plymouth, New Zealand.

- ■ highest risk: pyroclastic lava flows
- ■ high risk: lahars and flooding
- ■ low risk: lahars and flooding
- ■ very low risk
- ■ unlikely to be affected

Figure 1

> Think about how this data can be used to manage the future development of a country.

Exam focus

Most questions will have a resource linked to them. If the question asks you to use a figure, refer to specific details from the image in your answer.

Using **Figure 1**, suggest how hazard mapping can be used to help reduce the risk of tectonic hazards.   **[2 marks]**

⌐ Hazard mapping can be used to predict the location of future hazards, which allows ....

....

....

⑩ Monitoring and protection   Grade 4

2. Outline **one** protection strategy that can help reduce the risk of tectonic hazards.   **[2 marks]**

....

....

....

3. Outline how earthquake-proof buildings can reduce tectonic hazard risk.   **[2 marks]**

⌐ Earthquake-proof buildings can reduce the risk by ....

....

....

4. Explain how seismometers are important for monitoring volcanoes and earthquakes.   **[2 marks]**

....

....

....

General atmospheric circulation model

BBC

② Quick quiz

1. Which global circulation cell is found between 0° and 30°N? Tick **one** box.

☐ Polar cell

☐ Ferrel cell

☐ Hadley cell

2. What are the trade winds? Tick the correct definition.

☐ Cool air that moves back towards the equator

☐ Cool air that moves away from the equator

☐ Winds that blow near the poles

⑤ Global atmospheric circulation   Grade 3

1. Study **Figure 1**, which shows the cells involved in atmospheric circulation.

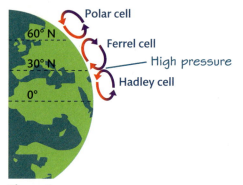

Figure 1

Label an area of low pressure and an area of high pressure.　**[2 marks]**

2. Using **Figure 1**, complete the following sentences.　**[2 marks]**

At latitudes of 60° north and south, the .... air of the Ferrel cells meets colder polar air at the .... fronts. A polar jet stream is formed above this, which drives the unstable atmosphere.

> Warm rising air causes low pressure and cool descending air causes high pressure.

Exam focus 📌

If you are asked to label a diagram, make sure you clearly identify where you are placing your label. Try practising drawing and labelling diagrams when you are revising. Check your diagrams against diagrams in your notes, textbooks or revision websites.

Geographical skills 🌐

Lines of latitude run horizontally around the Earth's surface from east to west. Make sure you are familiar with the key lines of latitude, such as the equator and the tropics, because you could be asked to refer to them when using a map.

⑤ Global weather and climate systems   Grade 5

3. Explain why hot deserts are found at 30° north and 30° south of the equator.　**[4 marks]**

....
....
....
....
....
....

Distribution of tropical storms

② Quick quiz

1. Complete the sentence by ticking **one** of the options below.

   Tropical storms tend to occur in the northern tropics between...

   ☐ ...June and November.   ☐ ...November and May.

3. Complete the sentence by circling the correct option.

   The Coriolis effect is **weaker** / **stronger** closer to the equator, so tropical cyclones do not form in this area.

2. Complete the sentence by ticking **one** of the options below.

   Cyclones begin in the...

   ☐ ...Atlantic Ocean.

   ☐ ...South Pacific or the Indian Ocean.

4. Complete the sentence.

   Tropical storms need a lot of .... in the atmosphere to form.

⑤ Distribution of tropical storms   Grade 3

1. Study **Figure 1**, a map showing where tropical storms develop.

Figure 1

Geographical skills

If you are asked to describe the track of tropical storms in your exam, use the following guide:
1. Name an ocean where they originated.
2. Identify the direction of movement using compass points.
3. Name a place where they may hit land.

Make sure you understand the relationship between the global circulation system and the conditions needed for tropical storms. Tropical storms tend to move westwards, because they are affected by the prevailing trade winds. The Coriolis effect must create spin for them to form, so they cannot form at the equator.

Using **Figure 1**, describe the track of hurricanes.   **[2 marks]**

Hurricanes can form in the Pacific and .... oceans, ...
moving west. Hurricanes in the Atlantic Ocean move towards ....
....

2. Using **Figure 1**, describe the track of typhoons.   **[2 marks]**

....

⑤ Conditions needed for tropical storms   Grade 3

3. Give **two** conditions that are needed for a tropical storm to form.   **[2 marks]**

   ....
   ....
   ....

4. Explain why tropical storms occur between the tropics.   **[2 marks]**

   ....
   ....
   ....

 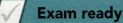

Causes and features of tropical storms

② Quick quiz

1. Complete the sentence.

In a tropical storm, the .... the pressure, the faster the wind and the stronger the storm.

2. Which type of cloud is formed when water evaporates from the warm sea surface?

....

⑤ Features of tropical storms   Grade 3

1. Study **Figure 1**, a diagram showing features of a tropical storm.

Figure 1

Using the labels below, complete **Figure 1**.   **[2 marks]**

warm, moist air   strong winds   eye

2. Which of the following statements are true about the features of tropical storms? Shade **two** circles only.   **[2 marks]**

A The eye of the storm is a region of clear skies where cool air is descending.   ◯

B The weakest winds tend to occur in the eye wall.   ◯

C Tropical storms create very low pressure, high winds, and heavy rainfall.   ◯

D Tropical storms create very high pressure.   ◯

⑤ Global weather and climate systems   Grade 5

3. Explain how climate change can affect tropical storms.   **[2 marks]**

🪧 Rising sea temperatures from climate change can affect the ....

distribution of tropical storms ....

....

....

> This question is worth 2 marks so you could give two ways that climate change can affect tropical storms or develop one point.

4. Complete the following sentences.   **[2 marks]**

A tropical storm is formed when thunderstorms converge, creating an area of very .... pressure. Air around this area .... and spins very fast.

> Make sure you understand all the main features of a tropical storm, where they are and what processes are taking place. For example, there are clear skies in the eye of the storm. You might be asked to label a diagram or describe a tropical storm in your exam.

 Made a start   **Feeling confident**   ✓ **Exam ready**

Named example: Tropical storms

② Quick quiz

1. Complete the sentences.

   (a) Tropical storms can cause storm ...., which can lead to widespread coastal flooding.

   (b) .... are the direct impacts of a tropical storm.

⑤ Effects of tropical storms — Grade 3

1. Study **Figure 1**, a photograph showing damage to beachfront property in an area affected by a tropical storm.

   Suggest **one** other primary effect of a tropical storm.
   [2 marks]

   ....

   ....

Figure 1

2. Using **Figure 1**, describe **two** possible secondary effects of a tropical storm. **[2 marks]**

   Effect 1 Destruction to homes leaving people

   homeless

   Effect 2 ....

   ....

Exam focus

For a short-answer question using the command word **describe**, you will need to give the main points or characteristics. Be sure to note how many points you are asked to describe.

Secondary effects are the indirect impacts. They are often long-term, for example homelessness.

⑩ Responses to tropical storms — Grade 4

3. Using an example you have studied, outline **one** long-term response to a tropical storm that a government could use to support people affected by the storm. **[2 marks]**

   ....

   ....

4. Using an example you have studied, outline **one** immediate response to a tropical storm from foreign governments. **[2 marks]**

   ....

   ....

5. Explain how non-governmental organisations (NGOs) can provide support to countries in responding to a tropical storm. **[2 marks]**

   ....

   ....

Named example

You need to know a named example of a tropical storm to show its effects and responses. You could use it to support your answers to questions **3**, **4** and **5**.

Planning for tropical storms

② Quick quiz

1. State whether each image is an example of monitoring, prediction, protection or planning for tropical storms.

Figure 1

....

Figure 2

....

⑤ Planning and protection strategies

Grade 4

1. Give **two** planning strategies which can be used to manage the risk from tropical storms. **[2 marks]**

   Strategy 1 Advising people how to prepare for and respond to a tropical storm.

   ....

   Strategy 2 ....

   ....

2. Outline **one** way countries can protect themselves from tropical storms. **[2 marks]**

   Countries can construct storm shelters, which include features

   that make them strong enough to survive tropical storms.

   For example, ...

> **Exam focus** 📌
>
> Where there are two separate spaces for responses, write a different response in each space provided.

> Make sure you understand the difference between monitoring, prediction, protection and planning strategies to reduce the hazard risk of tropical storms. Learn examples of each type of strategy. For example, protection strategies include building storm shelters and constructing sea walls.

⑩ Monitoring tropical storms

Grade 5

3. Explain the benefits of monitoring tropical storms. **[4 marks]**

   ....

   ....

   ....

   ....

4. Suggest why lower income countries (LICs) are often less prepared for tropical storms than higher income countries (HICs). **[2 marks]**

   ....

   ....

 Made a start   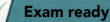 **Feeling confident**   **Exam ready**

Weather in the UK

---

② Quick quiz

1. Complete the sentence by ticking **one** of the options below.

   Precipitation is usually measured in...

   ☐ ...millimetres (mm) or metres (m)

   ☐ ...millilitres (ml) or litres (l)

2. List **three** types of extreme weather experienced in the UK.

   ....

   ....

   ....

---

⑤ Risk of flooding in London   Grade 3

1. Study **Figure 1**, which shows the number of properties at risk of flooding in London boroughs. Describe the distribution of properties at risk of flooding in London.   **[2 marks]**

   🚩 *The map shows that properties in the centre are*

   at a greater risk of flooding. For example,

   ....

   ....

2. Using **Figure 1**, which **one** of the following statements is true?   **[1 mark]**

   A There are fewer than 50 000 properties at risk of flooding in Hammersmith and Fulham.   ◯

   B Most of the boroughs in the South of London have more than 50 000 properties at risk of flooding.   ◯

   C There are fewer than 30 000 properties at risk of flooding in Harrow.   ◯

   D There are more properties at risk of flooding in Newham than in Wandsworth.   ◯

Harrow Newham

Hammersmith and Fulham Southwark

Wandsworth

Key
Number of properties at risk

■ 50 000+   ■ 30 001–40 000   ■ 20 001–30 000

■ 10 001–20 000   ■ 0–10 000

Figure 1 London properties at risk of flooding by borough

Geographical skills 🌐

Figure 1 is a choropleth map. It uses colour to show how data varies over geographical areas.

---

② Extreme weather in the UK   Grade 3

3. Study **Figure 2**, a map of the UK showing temperatures forecast in July 2015.

   Complete the following sentences.   **[2 marks]**

   The highest temperature forecast in July 2015 was in the .... region of the UK.

   Heatwaves can cause long periods of .... leading to extra pressure on water resources.

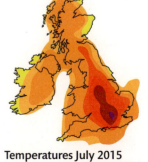

Temperatures July 2015

10°C ▬▬▬▬▬▬ 30°C+

Figure 2

 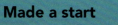 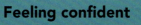

UK extreme weather events

② **Quick quiz**

1. Is the statement below true or false? Circle the correct answer.

Average temperatures in the UK were higher in 2016 than the average temperatures during the 20th century. This suggests that weather in the UK is becoming more extreme.

true   false

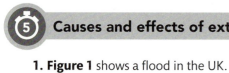
⑤ **Causes and effects of extreme weather events**   **Grade 3**

1. Figure 1 shows a flood in the UK.

Using **Figure 1**, suggest **one** social and **one** economic impact of the flood.

[2 marks]

....

....

....

Figure 1

2. Outline a possible cause of an extreme weather event.   **[2 marks]**

🚩 An example of an extreme weather event is .... This can be caused by ....

....

⑤ **Management and impacts of extreme weather**   **Grades 5–9**

3. Suggest a possible management strategy that could help to reduce the risk of flooding in the UK.   **[2 marks]**

....

....

....

4. Using a named example you have studied, explain the economic and social impacts of an extreme weather event in the UK.   **[6 marks]**

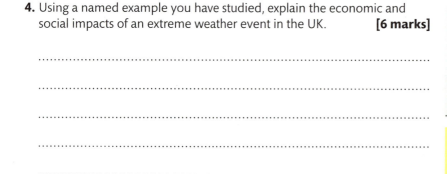

📍 **Named example**

You need to know an example of a recent extreme weather event in the UK. Revise the causes, impacts, and the management strategies that were used for your example.

For a 6-mark question you should support your answer with facts specific to the extreme weather event you have studied.

....

....

....

....

Climate change evidence

② Quick quiz

1. Is this statement true or false? Circle the correct answer.

Climate change could affect food supply by allowing more crops to be grown in some areas and reducing the crops that can be grown in other places.

true   false

2. Name **one** method that scientists can use to find out what the climate was like thousands of years ago.

....

⑤ Evidence for climate change   Grade 3

1. Give **one** piece of evidence that demonstrates climate change.

[2 marks]

....

....

....

> **Named example** 📍
>
> You could write about a named example. For example, glaciers in the Alps and the Himalayas.

2. Complete the sentences below.   **[2 marks]**

Tree rings tend to grow .... in warm, wet years and thinner in cold, dry years. By compiling

tree ring data, scientists can gather evidence about how the .... has changed over time.

⑤ Sea-level change   Grade 3

3. Study **Figure 1**, which shows sea level change between 1993 and 2015. Give the change in sea level in 2013.

[1 mark]

....

....

> **Exam focus** 📌
>
> If you need to use data from a graph in your exam, use a ruler to obtain an exact figure.

4. Give the change in sea level in 2001 shown in **Figure 1**.

[1 mark]

....

....

5. Describe the trend in sea level change for the period shown in **Figure 1**.   **[2 marks]**

Global sea level change in mm

```
60
40
20
 0
-20
    1995 1998 2001 2004 2007 2010 2013
                    Year
```

Figure 1

....

....

....

....

....

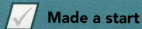

Causes of climate change

② Quick quiz

1. Are these statements true or false? Circle the correct answers.

(a) The greenhouse effect refers to heat from the Sun being trapped in the Earth's atmosphere by greenhouse gases.

true   false

(b) Climate change is only caused by natural factors.   true   false

2. Name **two** fossil fuels. ....

⑤ Causes of climate change   Grade 3 ☑

1. Outline how livestock farming can contribute to climate change.

[2 marks]

The farming of livestock, especially cows, causes the release of

methane, which ....

....

....

2. Outline how volcanic activity can cause global warming.   **[2 marks]**

....

....

....

> **Exam focus**
>
> For this **outline** question, you need to set out the main characteristics of how livestock farming can affect climate change.

> Question **1** only asks about livestock farming, but you also need to know how other types of agriculture cause climate change. For example, rice farming produces methane, a greenhouse gas.

> Volcanic activity can cause both global warming and global cooling.

⑤ Orbital changes   Grade 4 ☑

3. Give **two** reasons why orbital changes affect the Earth's climate.   **[2 marks]**

Reason 1 ....

....

Reason 2 ....

....

⑤ Human causes of climate change   Grade 4 ☑

4. Explain how human activity is contributing to climate change.   **[2 marks]**

....

....

....

 Made a start   **Feeling confident**   **Exam ready**

Managing climate change

② Quick quiz

1. Complete the sentence by ticking **one** of the options below.

Reducing the causes of climate change is known as...

☐ ...adaptation.

☐ ...mitigation.

2. Complete the sentence by ticking **one** of the options below.

Carbon capture is a type of...

☐ ...mitigation.

☐ ...adaptation.

⑤ Adaptation and mitigation   Grade 3

1. Outline **one** way adaptation could reduce the effects of climate change.
[2 marks]

....
....
....
....

Exam focus

You will have learned about several adaptation to climate change in your course. If you are asked to write about one of them, choose the one you are most confident writing about.

2. Explain why international agreements can contribute to managing climate change. **[2 marks]**

International agreements, such as the 2015 Paris Agreement, ...
....
....
....

⑤ Planting trees and changes to agriculture   Grade 4

3. Outline how planting trees can contribute to mitigating climate change. **[2 marks]**

....
....
....
....

4. Explain how agricultural systems can be adapted to climate change.
[3 marks]

Exam focus

This **explain** question is worth 3 marks so include three points.

....
....
....
....
....

Using examples and case studies

 15 **Living in hazardous areas** **Grades 5–9**

1. Choose **either** an earthquake **or** a volcanic eruption.
Assess the extent to which immediate responses are more significant than long-term responses.
Use an example you have studied.

[9 marks]
[+3 SPaG marks]

For this type of 9-mark question, you will need to use your own knowledge of examples (AO1 3 marks) and your own understanding to explain the role of immediate and long-term responses (AO2 3 marks) and use evidence to make a reasoned judgement (AO3 3 marks) about which you think are most important.

Think about using the following terms when making your judgement: most important, essential, vital, considerable, significant, hugely important, slight, limited.

Exam focus

Refer to named examples and case studies even if the question does not ask for it, especially in your answers to 6- and 9-mark questions. You should demonstrate detailed knowledge of your case studies and examples, and use specific facts to support your answer.

Both immediate and long-term responses are significant when reacting to an earthquake or volcanic

eruption and can, to a large extent, determine how severe the impact is on the environment and on

people in the area affected by the hazard.

....

....

....

....

....

....

....

....

....   Continue your answer on your own paper.

 10 **Climate change** **Grades 5–9**

2. Use evidence to support this statement.   **[6 marks]**

'Human activity is increasing the speed and severity of climate change.'

Named example

If you are asked to use evidence in your answer, you must include facts and information from an example or case study. This is essential for accessing the top marks.

....

....

....

....

....   Continue your answer on your own paper.

 Made a start   **Feeling confident**   **Exam ready**

A small-scale ecosystem

⑤ Quick quiz

1. Is this statement true or false? Circle the correct answer.

   The distribution of the Earth's large-scale global ecosystems, such as temperate forest, grassland and tropical rainforest, is not affected by climate.

   true   false

2. Complete the sentence by ticking **one** of the options below.
   The abiotic components of an ecosystem are...

   ☐ ...all the living things.

   ☐ ...all the non-living things.

4. Name **two** global ecosystems.

   ....

   ....

3. What is a carnivore?

   ....

⑩ Ecosystem interactions   Grades 2–3

1. Study **Figure 1**, a diagram showing a pond ecosystem. What is the name given to organisms like the pickleweed and algae?   **[1 mark]**

   ....

2. Identify a primary consumer shown in **Figure 1**.   **[1 mark]**

   ....

3. Define the term 'food web'.   **[1 mark]**

    A food web shows how all the food chains

   ....

4. Suggest how a change in the population of carnivores will impact on the ecosystem shown in **Figure 1**.   **[2 marks]**

   ....

   ....

   ....

   ....

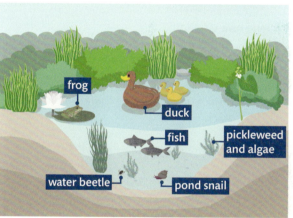

Figure 1

> Primary consumers are herbivores (animals that feed on plants).

Named example

Try revising this topic by drawing food webs for your named example location and classifying the main organisms found there. For example, 'algae – producer'. This can help you learn the interrelationships in your example ecosystem.

② Ecosystem producers   Grade 4

5. Identify which statement correctly describes the role of producers in an ecosystem. Shade **one** circle only.   **[1 mark]**

   A Fungi and bacteria break down organic material.   ◯

   B Organisms generate energy from feeding on other organisms.   ◯

   C Organisms produce their own food by converting sunlight.   ◯

Exam focus

For multiple-choice questions, read through each option carefully before you make your decision. You won't be awarded the mark if you shade the wrong number of circles.

 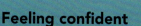

Tropical rainforests

② Quick quiz

1. Name the layer of vegetation that is nearest the ground in a tropical rainforest. ....

2. Complete the following sentences.

   (a) Tropical rainforest climates are .... and .... all year round.

   (b) Plants have adaptations such as .... leaves and .... roots to help them survive.

② Distribution of tropical rainforests   Grade 2

1. Study **Figure 1**, a map of rainforests of the world. Which **one** of the following statements is correct?

   [1 mark]

   Shade **one** circle only.

   A Large areas of tropical rainforest are found in South America, North America and Europe. ○

   B Most tropical rainforests are found within the tropics. ○

   C Tropical rainforests are not found in Africa. ○

   D The largest area of tropical rainforests is in Australia. ○

Atlantic Ocean · Asia · Tropic of Cancer · Africa · South America · Indian Ocean · Pacific Ocean · Australia · Tropic of Capricorn · Madagascar

■ Tropical rainforests of the world

Figure 1

Exam focus

If you are given a map in the exam, make sure you study it carefully, even if you think you already know the answer.

⑩ Characteristics of tropical rainforests   Grade 4

2. Explain why tropical rainforests have very high levels of biodiversity. **[2 marks]**

   The climate in tropical rainforests is always hot and wet, which means the growing season lasts ....

   ....

   ....

3. Outline **two** characteristics of tropical rainforests. **[4 marks]**

   ....

   ....

   ....

   ....

   ....

   ....

   ....

Exam focus

This question is worth 4 marks, so briefly develop both of your points.

 Made a start   Feeling confident   ☑ Exam ready

Case study: Deforestation

② Quick quiz

1. Give **one** environmental impact of deforestation.

....

....

2. Name **three** causes of deforestation.

....

....

....

⑤ Plate margins   Grade 3

1. Study **Figure 1**, a graph showing tree cover loss in Indonesia's forests.

(a) Which year had the greatest tree cover loss?

[1 mark]

....

(b) State the annual tree cover loss in 2004. **[1 mark]**

....

(c) Describe the overall trend in tree cover loss in Indonesia's rainforests. **[2 marks]**

📍 *Figure 1 shows an overall increase in the rate of*

deforestation and the hectares of tree cover lost

....

....

Annual tree cover loss in millions of hectares vs Year (2001–2014)

Figure 1

Geographical skills 🌐

If you are asked to describe a trend, you should include the following in your answer:

1. What the trend is – increase or decrease.

2. What the change has been – you should include specific numbers from the figure you have been given.

⑤ Causes of deforestation   Grade 4

2. Explain how agriculture can cause deforestation.

[4 marks]

....

....

....

....

....

....

....

....

....

Explain the cause and effect between agricultural methods and deforestation. After you have described a type of agriculture, use a phrase such as, 'This can lead to deforestation because…' to explain your answer.

Case study 🔍

Revise the causes and impacts of deforestation for your tropical rainforest case study. Use facts to support your answers in the exam.

Sustainable rainforests

② Quick quiz

1. Complete the sentences using the words below.

(a) Tropical rainforests are valuable to the environment because of their very high level of ....

(b) .... conserve and provide education about particular areas.

(c) Debt .... involves writing off some of a country's debt if they commit to the .... of their rainforest.

| national parks | biodiversity | protection | urban areas | moisture | reduction |

⑤ Ecotourism   Grade 4

1. Study **Figure 1**, a photo showing people soft trekking in Indonesia.

Outline the benefits of soft trekking as an approach to sustainable rainforest management.   **[2 marks]**

⌐ Soft trekking guides tourists on a specifically allocated trail,.........

which prevents ....

....

....

Figure 1

2. Give **one** reason why ecotourism is important to sustainable rainforest management.   **[1 mark]**

....

....

⑤ Sustainable management strategies   Grade 4

3. Explain the importance of selective logging to the preservation of tropical rainforests.   **[3 marks]**

....

....

....

....

....

....

> The key word is 'selective' – logging that isn't selective is a major cause of deforestation.

> You could also be asked about the importance of tree replanting, conservation and education, ecotourism, debt reduction and international agreements.

4. State **one** aim of international co-operation in the management of rainforests.   **[1 mark]**

....

....

✓ **Made a start**   ✓ **Feeling confident**   ✓ **Exam ready**

Characteristics of hot deserts

② Quick quiz

1. Complete the sentence by ticking **one** of the options below.

   Deserts receive less than .... rainfall per year.

   ☐ 150 mm   ☐ 250 mm   ☐ 350 mm

2. Complete the sentence by ticking **one** of the options below.

   Hot deserts cover approximately ....
   of the Earth's land surface.

   ☐ 0%   ☐ 40%   ☐ 20%

3. Name the largest desert in the world.

   ....

4. Name **one** threat to freshwater oases.

   ....

⑩ Characteristics of hot deserts   Grades 1–2 ☑

1. Choose the statement below that describes the climate of hot deserts. Shade **one** circle only.   **[1 mark]**

   A Hot during the day with temperatures reaching over 50 °C, but cold night-time temperatures which can drop below freezing.   ◯

   B Mild temperatures most of the year (10–20 °C).   ◯

   C Seasonal temperatures during the year (10–30 °C).   ◯

   > **Exam focus**
   >
   > **Outline** means 'set out the characteristics', so you do not need to include any explanations in your answers to these questions. Look at the number of marks to gauge how many points you need to make; this question is worth 2 marks, so outline two characteristics of hot deserts.

2. Outline the characteristics of hot deserts.   **[2 marks]**

   One of the characteristics of hot deserts is that they are very dry, with average precipitation levels

   of less than ....

   ....

   ....

⑤ Adaptations   Grades 1–2 ☑

3. Explain **one** way animals have adapted to hot deserts.   **[2 marks]**

   Camels have adapted to surviving without water for several

   ....

   ....

4. Outline **one** way plants have adapted to survive in hot deserts.   **[2 marks]**

   ....

   ....

   ....

Case study: Opportunities and challenges in a hot desert

② Quick quiz

1. Complete the sentence.

   Limestone, copper and phosphate are all .... that can be found in hot deserts.

2. Quad biking tours are an example of which opportunity provided by hot deserts?

   ☐ farming   ☐ tourism

3. Growing dates and other fruit is an example of which opportunity provided by hot deserts?

   ....

4. Complete the sentence.

   Due to their consistent levels of sunlight, deserts provide the opportunity to generate .... energy.

⑤ Challenges of hot deserts   Grade 4

1. Explain **two** challenges of living in hot deserts.   **[4 marks]**

   🚩 The harsh landscape of deserts such as the Sahara in North ....

   Africa make it difficult for people to develop settlements ....

   because of difficulties building key infrastructure. ....

   Another challenge ....

   ....

   ....

   ....

   ....

> Think about the physical characteristics of deserts that make them difficult for humans to survive in.

> **Case study**
> Use information from your case study to support your answers, even if the question does not specifically ask for it. If you do not remember specific details of the case study, just name an example and its location to help focus your answer.

⑩ Development opportunities in hot deserts   Grades 5–9

2. 'Hot deserts provide economic opportunities.' Do you agree with this? Using your understanding, explain your answer.   **[6 marks]**

   ....

   ....

   ....

   ....

   ....

   ....

   ....

✓ **Made a start**   ✓ **Feeling confident**   ✓ **Exam ready**

Desertification

② Quick quiz

1. Complete the sentence.

Areas on the edge of .... are particularly at risk of desertification.

2. Give **one** advantage of small-scale irrigation systems.

....

3. Name the cause of desertification being described in each sentence below.

(a) Sheep, cattle and goats eat too much vegetation, leaving the soil exposed.

....

(b) People cut down trees for fuel, causing the roots to die and removing the binding agent for the soil.

....

⑤ Causes of desertification
Grades 2–3

1. Study **Figure 1**, a photo showing a farmer in the Sahel region of Africa.
Outline how over-cultivation can cause desertification. **[2 marks]**

🪧 The intensive growing of crops to meet the

demands of growing populations means that the

soil is unable

....

....

....

Figure 1

> Think about what crops need from the soil to be able to grow.

> Question **1** asks **how** over-cultivation can lead to desertification. You need to give an explanation, clearly showing the relationship between over-cultivation and desertification.

2. Outline how climate change is causing desertification. **[2 marks]**

....

....

....

⑤ Managing desertification
Grade 4

3. Explain how the use of appropriate technology can help to reduce desertification. **[2 marks]**

....

....

....

....

....

....

Characteristics of cold environments

② Quick quiz

1. Where is the Arctic Tundra biome located? Circle the correct answer.

   Northern Hemisphere   Southern Hemisphere

2. Name **two** animals that are adapted to living in cold environments.

   ....

   ....

3. Complete the sentence by ticking **one** of the options below.

   Polar regions cover the Arctic and...   ☐ Asia   ☐ Antarctic   ☐ Africa

⑩ Adaptations and biodiversity   Grade 3

1. Study **Figure 1**, a photo showing the Labrador tea plant growing in Alaska. Its leaves have fine hairs on their underside to help prevent water loss and retain heat.

   Using **Figure 1**, explain **one** other way plants have adapted to the cold environment.   **[2 marks]**

   🪧 In Figure 1, you can see that mosses and lichens grow close

   ....

   ....

   ....

Figure 1

You could write about a different element of the Labrador tea plant or another plant shown in the photo, such as mosses.

2. Explain why cold environments have low biodiversity.   **[2 marks]**

   ....

   ....

3. Complete the following sentences.   **[2 marks]**

   Plants have adapted to the extreme climate by growing quickly in .... when temperatures rise. They provide a .... for ground nesting birds and are food for herbivores such as reindeer. This is an example of interdependence.

⑤ Characteristics of cold environments   Grade 4

4. Outline the characteristics of cold environments.   **[3 marks]**

   ....

   ....

   ....

   ....

   Exam focus 📌

   This question is worth 3 marks, so outline three characteristics.

 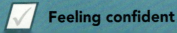

Case study: Opportunities and challenges in tundra

② Quick quiz

1. Complete the sentence.

Buildings are built on stilts to protect them from flooding when the .... melts.

2. Name **two** possible development opportunities in cold environments.

....

⑤ Development opportunities   Grade 3

1. Figure 1 and **Figure 2** show challenges and opportunities in the Siberian tundra.

Exam focus 📌

Use the photos you are given as prompts for your answer, even if your case study is about a different area.

Figure 1   **Figure 2**

(a) Describe **one** opportunity for development in a cold environment.   **[2 marks]**

In Siberia, the landscape and wildlife ....

....

(b) Describe **one** way in which the climate of a cold environment can be a challenge for development.   **[2 marks]**

....

....

⑤ Challenges in cold environments   Grade 4

2. Identify the statement below that represents a challenge of developing cold environments. Shade **one** circle only.   **[1 mark]**

A Oil fields allow for extraction and production of crude oil.   ◯

B Many cold environments have an abundance of minerals.   ◯

C Many cold environments are extremely remote and only accessible by reindeer or snowmobile.   ◯

Make sure you read through all the options in a multiple-choice question carefully.

3. Outline how extreme temperatures can create challenges for people living in cold environments.   **[2 marks]**

....

....

....

....

....

....

Conservation of cold environments

⏱ Quick quiz

1. Complete the sentence.

Conserving cold environments by preventing ice sheets from .... is essential to managing the

global .... in sea levels.

2. State **one** role the World Wide Fund for Nature (WWF) plays in Antarctica. ....

3. Is this statement true or false? Circle your answer.

Governments are important for protecting cold environments because
they can pass laws that protect important areas from economic activity.

true   false

⏱ Importance of cold environments

Grade 4

1. Explain the importance of conserving cold environments as
wilderness areas. **[2 marks]**

🪧 Cold environments provide a home for many indigenous tribes ....

who rely on animals for their food, clothing and tools. If these .....

environments were not protected wilderness areas, ...

....

....

> You could write about other reasons for conservation, such as the habitat it provides for wildlife or development opportunities like leisure activities.

⏱ Managing cold environments

Grade 4

2. Outline the importance of international agreements in the management
of cold environments. **[2 marks]**

🪧 International agreements provide clear guidelines to establish ....

strategies and agreements to best protect the environment. ....

....

....

....

> You could use the example of the Antarctic Treaty, an agreement between 12 countries to protect the Antarctic environment.

3. Explain how the use of technology can be managed to preserve cold environments. **[3 marks]**

....

....

....

....

 Made a start   Feeling confident   ✓ Exam ready

Landscapes in the UK

② Quick quiz

1. Name the UK's highest mountain.

....

2. Name the UK's longest river.

....

3. In which region are most of the upland areas in the UK? Tick **one** answer.

☐ south-west   ☐ north-west

☐ south-east   ☐ north-east

4. Complete the sentence by ticking **one** of the options below.

The north of Scotland is a mostly .... area.

☐ flat   ☐ upland   ☐ lowland

⑤ Characteristics of upland and lowland landscapes   Grade 2

1. Outline why upland landscapes are suitable for sheep farming. **[2 marks]**

> Think about what soil conditions wouldn't be suitable for growing crops.

Upland landscapes are suitable for sheep farming because the

soil tends to be ....

....

....

2. State **one** characteristic of lowland landscapes.   **[1 mark]**

....

⑤ Ecosystem producers   Grade 4

3. Study **Figure 1**, which shows a map of Ben Nevis.

(a) Identify the highest point for Ben Nevis.
Shade **one** circle only.   **[1 mark]**

A 1300 m   ◯

B 1345 m   ◯

C 1320 m   ◯

(b) Give the four-figure grid reference for Ben Nevis.
Shade **one** circle only.   **[1 mark]**

A 1671   ◯

B 1672   ◯

C 1773   ◯

(c) A mountain rescue post is identified using this symbol: 📞
Give the six-figure grid reference for the mountain rescue post.   **[1 mark]**

....

Figure 1 An OS map extract of Ben Nevis.
Scale 1 : 50 000

Geographical skills

Contour lines indicate points of equal height above sea level. The closer they are together, the steeper the gradient of a slope. Contour lines further apart represent a gentle slope.

Waves and weathering

---

⑤ Quick quiz

1. Complete the sentence by ticking **one** of the options below.

   The breakdown of rocks by tree roots is a form of
   .... weathering.

   ☐ mechanical   ☐ chemical

2. Complete the sentence by ticking **one** of the options below.

   The fetch is...

   ☐ ...how far the wave has travelled.
   ☐ ...the power of the wind.

---

⑩ Wave characteristics   **Grade 3**

1. Study **Figure 1**, a diagram showing a destructive wave.

[   ]   [   ]

[   ]

Figure 1

Complete the diagram above by writing the correct label in each box. Choose from the labels below.   **[2 marks]**

| weak swash | strong backwash (beach eroded) | tall waves with short wavelength |

2. State **two** characteristics of constructive waves.   **[2 marks]**

🚩 Constructive waves have a strong swash ....

....

....

3. State **two** factors that affect the energy of a wave.   **[2 marks]**

....

....

....

---

⑤ Weathering   **Grade 4**

4. Explain how chemical weathering impacts on coastal landscapes.   **[2 marks]**

🚩 When acidic rainwater falls on rocks, over time

this leads to ....

....

....

5. Explain how mechanical weathering impacts on coastal landscapes.   **[2 marks]**

....

....

....

....

 Made a start   **Feeling confident**   ☐ **Exam ready**

Coastal erosion landforms

② Quick quiz

1. Complete the sentence by ticking **one** of the options below.

Hydraulic action is...

☐ ...waves compressing pockets of air in cracks in a cliff.

☐ ...the chemical reaction of rocks with the acidity of the seawater.

2. Write down these features in the order in which they are formed.

arch   fault / crack   stack   cave   ....

3. Draw lines to match the process to its definition.

Abrasion	The action of rocks colliding with each other, causing them to become smoother and rounded
Attrition	The action of rock particles carried by waves being hurled at a cliff, causing pieces to break off

⏱ 20 Formation of erosional landforms   Grades 4–9 ☑

1. Complete the following sentence.   **[2 marks]**

Headlands and .... form when there are alternating bands of .... rock and softer, less resistant rock.

2. Explain the processes involved in the formation of wave cut platforms.
                                                          [4 marks]

🪧 *Destructive waves attack the base of a cliff through the sheer*

force of the waves compressing air into small cracks within the

rock, a process known as hydraulic action.

....

....

....

....

....

> **Exam focus** 📌
>
> For these **explain** questions (2 and 3), you need to develop your answers to include the type of erosion taking place and the role that the geology plays.

> Structure your answer logically. Explain each stage in the order that it occurs. This will help you to include all the processes involved in your answer without forgetting any.

3. Explain the processes involved in the formation of a coastal stump.
                                                          [4 marks]

....

....

....

....

....

> **Exam focus** 📌
>
> Use accurate specialist terminology in your answers. For example, include names of processes, such as hydraulic action and abrasion, and coastal landforms, such as arches, stacks and stumps.

Mass movement and transportation

② Quick quiz

1. Complete the sentence.

.... is a transport process in which dissolved minerals are carried within seawater.

2. Complete the sentence.

Traction is the .... of large pebbles along the sea floor/bed.

3. Complete the sentence by ticking **one** of the options below.

Slumping is... ☐ ...the sudden movement of large volumes of rock. ☐ ...the rotational collapse of permeable rocks.

4. At what angle do the waves approach the beach in the process of longshore drift? Circle the correct answer.

oblique angle   right angle

⑤ Longshore drift   Grades 2–4

1. Study **Figure 1**, a diagram showing the process of longshore drift.

-- backwash
— swash
— prevailing wind

Figure 1

> **Exam focus**
>
> When exam questions refer to a figure, such as a graph or photograph, always take time to carefully look over the resource to ensure you understand what it shows.

(a) State what influences the direction of longshore drift.   **[1 mark]**

> For question **1(a)**, think about what is moving the wave towards the coastline.

....

(b) What part of the wave brings the sediment onto the beach? **[1 mark]**

....

⑤ Mass movement   Grades 2–4

2. Explain why rock falls occur.   **[2 marks]**

🚩 *Rock falls happen due to the movement of rock*

fragments under the influence of gravity.

....

....

....

....

3. Which **one** of the following statements describes sliding? Shade **one** circle only.   **[1 mark]**

A The transportation of eroded material along the coast in a zig-zag pattern. ◯

B The rolling of large pebbles along the sea floor. ◯

C The collapse of unsupported material above a coastal arch. ◯

D The sudden movement of large volumes of rock and soil along a zone of saturated soil. ◯

Coastal deposition landforms

② Quick quiz

1. Name a grass that can help stabilise sand dunes. ....

2. Complete the sentence.

A spit is formed from the build-up of sand and sediment as a result of .... drift.

3. Complete the sentence.

The freshwater lake that can form behind a bar is called a ....

4. Which type of wave creates sandy beaches? ....

⑮ Formation of deposition landforms — Grades 4–9

1. Explain how sand dunes are formed. **[4 marks]**

Sand dunes are formed from onshore winds blowing sediment to ....

the back of the beach. The sediment is initially deposited around an

obstruction, leading to the formation of embryo dunes.

....

....

....

....

> Think about the type of vegetation that colonises sand dunes and the effect that this has on the dunes.

2. Describe and explain the processes involved in the formation of the landform in **Figure 1**. **[4 marks]**

....

....

....

....

....

....

....

....

....

....

....

....

Figure 1

Exam focus 📌

Think about all the processes involved in the formation of the landform, and explain how each one shapes the landform. To access the top marks, you should structure your answer logically, providing a clear description of the processes and a clear explanation of the sequence for the formation of features.

Hard engineering

⏱ ② Quick quiz

1. Identify the coastal management technique shown in **Figure 1**.

![Figure 1]

Figure 1

....

2. Complete the sentences.

(a) One benefit of using rock armour is...

....

(b) One disadvantage of using gabions is...

....

⏱ ⑤ Groynes and rock armour   Grade 3 ✓

1. **Figure 2** shows the use of groynes to manage the weathering of a beach. Explain the disadvantages of using groynes on a stretch of coastline.
   [2 marks]

   🪧 Groynes can help to reduce the effects of longshore drift but ....

   they can deprive ....

   ....

   ....

![Figure 2 Groynes at Lepe in the New Forest]

Figure 2 Groynes at Lepe in the New Forest

> Consider the impact of installing groynes on other areas along the coastline.

2. Identify which statement correctly describes rock armour. Shade **one** circle only.
   [1 mark]

   A Large boulders used to reduce wave energy ◯

   B The dumping of sand or shingle onto the beach that has been dredged offshore ◯

   C Curved concrete structures designed to reflect the energy of the waves ◯

> Revise the purpose, advantages and disadvantages of all hard engineering coastal management techniques.

⏱ ⑤ Management strategies   Grade 4 ✓

3. Hard engineering strategies are most effective for managing coastal erosion. Do you agree with this? Using examples you have studied, explain your answer.
   [6 marks]

....

....

....

....

....

....

📍 **Named example**
Use specific examples in your answers if they are relevant.

> Continue your answer on your own paper.

✓ **Made a start**   ✓ **Feeling confident**   ✓ **Exam ready**

Soft engineering

② Quick quiz

1. Draw lines to match each soft engineering strategy to its definition.

Dune regeneration	The transfer of sediment from the lower to the upper beach
Beach reprofiling	The artificial creation of new dunes or the restoration of existing dunes
Managed retreat	Allowing the sea to flood in designated areas of the coast

2. State **one** advantage of using managed retreat to manage coastlines.

....

② Beach nourishment — Grades 1–3

1. Identify which statement correctly describes beach nourishment. Shade **one** circle only. **[1 mark]**

A Large boulders used to reduce wave energy ⬭

B Depositing sand or shingle onto a beach to replace eroded material ⬭

C Curved concrete structures designed to reflect the energy of the waves ⬭

⑤ Dune regeneration and managed retreat — Grades 4–5

2. Figure 1 shows sand dunes on a beach in North Devon. Explain **one** benefit of using dune regeneration to manage coastlines. **[2 marks]**

One benefit of using sand dune regeneration is it

can contribute towards maintaining a diverse natural

environment, which helps to support ....

....

....

....

....

Figure 1 Sand dunes at Braunton Burrows, North Devon

3. Explain **one** cost associated with using managed retreat to protect coastlines. **[2 marks]**

....

> When you are revising coastal management techniques, consider the social, economic and environmental impacts of different types of management – a technique such as managed retreat can have a positive environmental impact by creating new habitats, but may not be popular with people living in the area.

....

....

....

....

....

....

Coastal management

---

(2) **Quick quiz**

1. Name **one** industry that could be affected if the coast is not protected from erosion.

....

2. Name **one** coastal management strategy that can create wildlife habitat.

....

3. Complete the sentence.

.... engineering strategies tend to be cheaper in the short term than

.... engineering strategies.

---

(5) **Identifying coastal features using grid references** | **Grades 2–4**

1. Study **Figure 1**, a map of Walton-on-the-Naze in the UK.

(a) Using **Figure 1**, state the six-figure grid reference for the Naze tower. **[1 mark]**

....

Geographical skills 🌐

For a six-figure grid reference, imagine the grid squares are divided into ten equal sections horizontally and vertically.

(b) Give the straight line distance from the Naze tower to the train station. Give your answer in km. **[1 mark]**

.... km

Geographical skills 🌐

On a 1:50 000 OS map, every 2 cm on the map is equal to 1 km in the real world.

Figure 1 Part of a 1:50 000 OS map showing Walton-on-the-Naze

---

(10) **Reasons for coastal management** | **Grades 5–9**

2. Coastal management schemes are important in protecting the coastline from physical processes. Do you agree with this? Using examples you have studied, explain your answer. **[6 marks]**

....
....
....
....
....
....
....

Named example 📍

You need to know an example of a coastal management scheme in the UK. The answer to this page uses Walton-on-the-Naze. Learn key facts such as the location, why coastal management was needed, what strategies were used, and their impact.

 Made a start   **Feeling confident**   ✓ **Exam ready**

A river profile

② Quick quiz

1. Complete the sentence by ticking **one** of the options below.

A tributary is... ☐ ...a stream that joins the main river. ☐ ...where the river starts.

2. What name is given to where the river flows out into the sea? ....

3. Complete the sentence.

The watershed is ....

4. Complete the sentence by ticking **one** of the options below.

In the lower course of a river the river channel is... ☐ ...narrow and shallow. ☐ ...wide and deep.

⑩ Upper, middle and lower course   Grades 4–5

1. Study **Figure 1**, which illustrates a river's long and cross profile.

Exam focus 📌

If you are asked to study a figure, such as a photo or diagram, in your exam, refer to details from the figure in your answer.

For part **(b)** think about the two different types of erosion that take place in a river channel: one causes downwards erosion and one causes sideways erosion.

Figure 1

(a) Describe the cross profile of the river in its upper course.   **[1 mark]**

....

(b) Suggest **one** reason why the cross profile changes between **A** and **B**.   **[1 mark]**

....

(c) At **C**, there is little vertical erosion of the valley sides. Suggest **one** reason for this.   **[2 marks]**

Refer to the upper, middle and lower course of the river in the correct order to explain changes in the river downstream.

🪧 *Point C is in the lower course, where the increased volume of*

water means that ....

....

2. Which **one** of the following correctly describes the changes in a river's bedload from the upper to the lower course? Shade **one** circle only.   **[1 mark]**

A The bedload is mostly sediment in the upper course, and small and angular in the lower course.   ◯

B The bedload is small and rounded in the middle and lower course.   ◯

C The bedload is large and angular in the upper course, smaller in the middle course and fine sediment in the lower course.   ◯

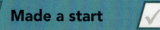

Fluvial processes

---

② Quick quiz

1. Which type of erosion is dominant at each stage of a river? Circle the correct answer(s).

(a) Upper course: vertical erosion   lateral erosion

(b) Middle course: vertical erosion   lateral erosion

(c) Lower course: vertical erosion   lateral erosion

2. Draw lines to match each process of erosion to its correct definition.

Abrasion	Soluble rocks, such as limestone, dissolve in the river
Attrition	Rock fragments carried by the river collide with one another, causing them to become smaller and more rounded
Solution	Rock particles carried by the river hit the bed and banks, wearing them down

---

⑤ Transportation and deposition   Grades 1–3

1. Identify which statement correctly describes the process of saltation. Shade **one** circle only.   **[1 mark]**

A The dissolving of rock types like limestone by the river   ○

B The bouncing of smaller pebbles along the river bed   ○

C The rolling of large pebbles along the river bed   ○

2. Identify which statement correctly describes the process of traction. Shade **one** circle only.   **[1 mark]**

A The dissolving of rock types like limestone by the river   ○

B The bouncing of smaller pebbles along the river bed   ○

C The rolling of large pebbles along the river bed   ○

3. Outline the process of deposition.   **[1 mark]**

....

....

4. State where deposition typically occurs along the course of a river.   **[1 mark]**

....

---

⑤ River erosion   Grades 2–4

5. Explain how the processes of hydraulic action and abrasion differ.   **[4 marks]**

Hydraulic action occurs when waves compress the air in cracks ......

in rocks, ...

....

....

....

....

> **Exam focus**
>
> Use comparative connectives, such as 'whereas', 'on the other hand' or 'however', to link your sentences together when describing differences between things.

      **Made a start**   **Feeling confident**   **Exam ready**

Fluvial erosion landforms

 Quick quiz

1. Which section of a river do waterfalls normally form in?

....

2. Name a fluvial erosion process that causes the formation of a waterfall.

....

 **Formation of waterfalls and gorges** **Grades 1–3**

1. Study **Figure 1**, which shows the High Force waterfall in Teesdale. State **one** difference between the types of rocks labelled **A** and **B**.

[1 mark]

....

....

Figure 1

2. Outline how a gorge can form downstream from a waterfall. **[2 marks]**

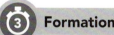 Gorges form when waterfalls retreat upstream.

This happens because ....

....

....

....

....

Look carefully at landform photos when answering formation questions, because they can give you clues on their formation. In **Figure 1**, you can see how the rocks are arranged differently, representing the difference in resistance.

Exam focus

To **outline** a process, you only need to give the main characteristics.

 Formation of interlocking spurs **Grades 3–4**

3. Study **Figure 2**, a photo of interlocking spurs in the Shropshire hills of England. Explain how interlocking spurs are formed. **[3 marks]**

....

....

....

....

....

....

Figure 2

Exam focus

To **explain** a process, you need to give the reasons for each stage happening.

Think about differences in rock type to explain how interlocking spurs are formed.

Fluvial erosion and deposition landforms

② Quick quiz

1. In what course of the river are meanders usually found? Circle the correct answer.

upper   middle   lower

2. Complete the sentence.

Meanders constantly move/migrate due to lateral erosion and ....

3. Are the statements below true or false? Circle the correct answer.

The fastest flow is on the outside bend of the meander.   true   false

A river cliff is formed from the deposition of materials.   true   false

A levée is an elevated bank along a river's edge.   true   false

⑩ Meanders   Grades 2–3

1. Study **Figure 1**, a diagram showing the cross profile of a meander.

(a) state the name of features **X** and **Y**.   **[2 marks]**

🚩 Feature X is a slip-off slope. ....

Feature Y is ....

Figure 1

(b) Using **Figure 1**, explain how a river cliff is formed.   **[3 marks]**

🚩 A river cliff is formed where the line of fastest flow ....

....

....

....

② Ox-bow lake formation   Grades 2–4

2. Study **Figure 2**, an aerial view of the formation of an ox-bow lake. Label an area of erosion and an area of deposition on the diagram.   **[2 marks]**

Figure 2

② Estuaries   Grades 2–4

3. Outline the characteristics of an estuary.   **[2 marks]**

....

....

....

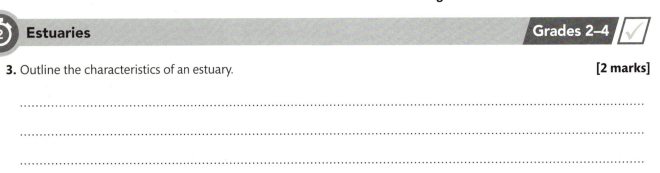
Made a start   Feeling confident   Exam ready

Flood risk

② Quick quiz

1. Are these factors social or economic? Circle the correct answer.

(a) The effects on people, for example the loss of life   social   economic

(b) The effects on the wealth of an area, for example the loss of business income   social   economic

2. Complete the sentence.

(a) Urbanisation creates lots of .... surfaces, which lead to increased surface run-off.

(b) Deforestation is a human factor that .... flood risk.

3. Does the statement below describe a physical factor or a human factor? Tick the correct answer.

Long periods of sustained, heavy precipitation result in the land being unable
to absorb more water and becoming saturated quickly.

☐ physical   ☐ human

⑩ Causes of flooding   Grades 2–3

1. Explain how relief affects flood risk.   **[2 marks]**

🚩 The steeper the surrounding slopes of an area,...

the greater the ....

....

....

....

> When answering an **explain** question, use connectives such as 'because' to show the examiner that you are giving a reason. You could also use 'this means that', 'therefore' or 'consequently'.

2. Outline **one** human factor which increases flood risk.
          [2 marks]

....

....

....

....

....

> You need to know all the human and physical factors that affect flood risk and how each factor increases or decreases the risk. For example, a sustained period of heavy precipitation is a physical factor that increases flood risk.

3. Which **one** of the following statements about how geology affects flood risk is true?
Shade **one** circle only.   **[1 mark]**

A Exposed soil causes more surface run-off, which reduces the risk of flooding.   ◯

B If the land near a river consists of permeable rock, the risk of flooding is lower because infiltration rates are higher, meaning water reaches the river channel at a slower rate.   ◯

C If the land near a river consists of impermeable rock, the risk of flooding is lower because infiltration rates are higher, meaning water reaches the river channel at a slower rate.   ◯

D Impermeable rocks, such as granite, have a greater rate of infiltration than permeable rocks, such as limestone.   ◯

Flood hydrographs

② Quick quiz

1. Complete the sentence.

A flood hydrograph shows the ways in which a
.... is affected by a ....

| discharge | river | storm | flood plain |

2. Complete the sentence.

A large drainage basin will cause ....
movement of water to river channels.

⑤ Features of a flood hydrograph — Grades 2–3

1. Study **Figure 1**, a flood hydrograph.

(a) Give the peak discharge in cumecs. **[1 mark]**

....

(b) Give the peak rainfall in mm. **[1 mark]**

....

(c) Calculate the lag time. **[1 mark]**

....

Geographical skills

If you need to obtain data from a flood hydrograph, use a ruler to make sure you get an accurate answer. For example, in order to answer question **1(a)** you need to use the right-hand scale and place your ruler at the top of the curve.

Figure 1

⑩ Factors affecting the shape of flood hydrographs — Grades 4–6

2. Describe the difference between a flashy and a gentle hydrograph. **[2 marks]**

....

....

....

3. Explain how the gradient of an area can affect the shape of a flood hydrograph. **[4 marks]**

....

....

....

....

....

....

....

Made a start ☐ Feeling confident ☐ Exam ready ☐

Hard engineering

② Quick quiz

1. Name the techniques described.

(a) This technique is where land on either side of a river is raised artificially. ....

(b) This technique involves the artificial removal of meanders from a river channel. ....

⑩ Causes of flooding

Grades 2–3

1. Study **Figure 1**, a photo of a dam and reservoir.

> When you're describing the costs, consider who might be relying on the sediments transported by the river for their livelihood.

> When you're describing the benefits, think about how the dam can be used to reduce reliance on the use of fossil fuels.

Figure 1

Explain the costs and benefits of using dams and reservoirs to manage rivers. **[4 marks]**

The use of dams and reservoirs can be effective in providing a source of water. Another benefit

is that they can be used to generate ....

....

....

....

....

② Flood relief channels

Grade 4

2. Explain why flood relief channels are used to manage rivers. **[2 marks]**

....

....

② Straightening

Grade 3

3. Outline the benefits of using straightening to manage river channels. **[2 marks]**

....

....

Soft engineering

② Quick quiz

1. Which soft engineering technique does each statement refer to? Tick the correct answer.

(a) This technique gives people time to move assets to a safer location.

☐ afforestation ☐ flood warnings and preparation

(b) This technique can decrease the value of land.

☐ river restoration ☐ flood plain zoning

⑤ Flood plain strategies   Grades 4–6 ☑

1. Identify which statement correctly describes flood plain zoning. Shade **one** circle only. **[1 mark]**

> Even if you think you know the answer, read through all the multiple-choice options carefully to make sure you choose the correct statement.

A The removal of any hard engineering strategies ◯

B The use of policies to control how land is used on, or near, flood plains ◯

C The planting of trees near the river channel ◯

2. Give **one** advantage of using flood plain zoning to manage flooding. **[1 mark]**

Flood plain zoning can provide land for alternative uses, such as ....

....

⑤ Costs and benefits   Grades 4–6 ☑

3. Explain **one** benefit of using river restoration to manage rivers. **[2 marks]**

> These questions ask for one cost or benefit, but make sure you revise both the costs **and** the benefits of all the river management strategies you have learned about.

....

....

....

....

....

4. Explain **one** cost associated with using afforestation to manage rivers. **[2 marks]**

....

....

....

....

....

 Made a start   ☑ **Feeling confident**   ☑ **Exam ready**

Flood management

② **Quick quiz**

1. Give **one** environmental effect of flooding.

2. Give **one** social effect of flooding.

....

....

⑩ **Impacts and issues**   Grades 2–3

1. Study **Figure 1**, a photo of the Boscastle floods in 2004. Describe the possible short-term impacts of this flood. **[2 marks]**

🚩 *One possible short-term impact of the flooding is the damage to properties. A second possible impact* ....

....

2. Suggest **one** possible economic effect of the river flood shown in **Figure 1**. **[2 marks]**

Figure 1

....

....

3. Explain **one** environmental issue associated with flood management schemes. **[2 marks]**

....

....

⑩ **Responses to flooding**   Grades 5–9

4. Flood management schemes are essential to protect people and places from the effects of river flooding. Do you agree with this? Using an example you have studied, explain your answer. **[6 marks]**

....

....

....

....

....

....

....

....

....

....

Named example 📍

For your exam, you need to know an example of a flood management scheme in the UK. Make sure you understand the social, economic and environmental issues associated with your case study.

You need to know the difference between cause, effect and response.
Cause: the reasons why the flooding event happened
Effect: the impacts the flooding had on people and the environment
Response: the strategies used to recover from the impacts of the flooding event

Glacial processes

⏱ ② Quick quiz ☑

1. Complete the sentence.

Bulldozing is the action of the the snout of a glacier

....

2. Is the statement below true or false? Circle the correct answer.

Rock fragments left in an unsorted pattern by a retreating glacier are known as glacial till.

true   false

3. What was the southernmost point of ice coverage reached in the UK in the last Ice Age?

....

4. What is the name given to the scratches left behind from the process of abrasion?

....

⏱ ⑤ Freeze-thaw weathering   Grades 1–3 ☑

1. Explain how freeze-thaw weathering can break down rocks over time. **[2 marks]**

> Remember that the end product of this process is the formation of rock fragments, known as scree.

🚩 Freeze-thaw weathering occurs when water

enters the cracks of rocks. When it freezes,..........

....

....

Exam focus 📌

Create a glossary of the key processes in the physical topics you are studying. It is important you can confidently recall, describe and explain these processes in your answers to exam questions.

⏱ ⑩ Glacial transportation and deposition   Grades 1–3 ☑

2. Give the definition of outwash. **[1 mark]**

....

3. Outline the transportation process of rotational slip. **[2 marks]**

....

....

4. Describe where glacial deposition mainly occurs. **[2 marks]**

....

....

⏱ ② Erosional processes   Grades 1–3 ☑

5. Identify which statement correctly describes the process of plucking. Shade **one** circle only. **[1 mark]**

A The sliding movement of the glacier from summer meltwater   ◯

B The cracking of the rock from repeated freezing and thawing   ◯

C Parts of the bedrock becoming attached to the glacier by freezing, and rocks being pulled out as the glacier continues moving   ◯

 Made a start   **Feeling confident**   **Exam ready**

Glacial erosion landforms

② Quick quiz

1. What are arêtes?

....

....

2. Complete the sentence.

A corrie forms from the accumulation of snow in a

.... on the side of a mountain.

3. Is the statement below true or false?
Circle the correct answer.

Truncated spurs form when a river cuts
straight through interlocking spurs.

true   false

4. Complete the sentence.

Pyramidal .... form when

several corries and .... meet,

and erosion results in the formation of a single peak.

⑤ Impacts and issues   Grades 2–3

1. Look at **Figure 1**, which shows the Snowdonia area of
Wales. Give the glacial landform in grid reference 633564.
Shade **one** circle only.   **[1 mark]**

A Arête   ○

B Glacial trough   ○

C Corrie   ○

2. Give the glacial landform in grid reference 626553.
Shade **one** circle only.   **[1 mark]**

A Arête   ○

B Glacial trough   ○

C Corrie   ○

Figure 1 An extract from an OS map of Snowdonia.
Scale 1:50000

3. Using **Figure 1**, identify the six-figure grid reference
for the Summit Station. Shade **one** circle only. **[1 mark]**

A 611552   ○

B 608561   ○

C 609544   ○

4. State the glacial landform in grid reference 617546 of
Figure 1. Shade **one** circle only.   **[1 mark]**

A Arête   ○

B Hanging valley   ○

C Corrie   ○

⑤ Formation of ribbon lakes   Grade 5

5. Explain how ribbon lakes are formed.   **[3 marks]**

....

....

....

....

Named example

Use your named example of erosion
and deposition landforms in an upland
area. For example, you could refer to
Ullswater in the Lake District as an
example of a ribbon lake.

Glacial transportation and deposition landforms

② Quick quiz

Draw lines to match the following key terms to the definition.

1 Lateral moraine		A Material that gathers down the middle of a glacier
2 Medial moraine		B Large boulders transported and deposited by a glacier
3 Erratics		C A series of ridges formed from the gathering of material behind the terminal moraine
4 Drumlins		D Material that gathers at the edge of a glacier due to freeze-thaw weathering
5 Recessional moraine		E Egg-shaped raised areas of land formed from moraine, found in clusters

⑤ Depositional features　　　　　　　　　　Grade 4

1. **(a)** Study **Figure 1**, a diagram showing different types of glacial moraine. Identify the type of moraine at **X**.　　**[1 mark]**

   ....

   (b) Describe this type of moraine.　　**[2 marks]**

   ....

   ....

   ....

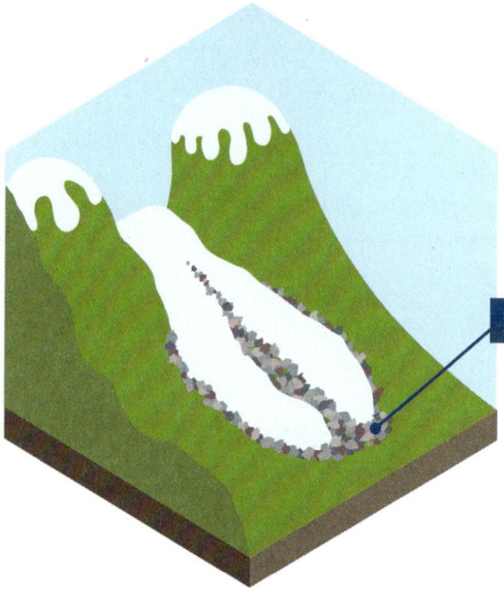

> Practise sketching glacial diagrams similar to **Figure 1**. Annotate them with as much information about the features shown as you can. Then check your diagram and annotations against your notes, a textbook or the internet.

Figure 1

⑤ Erratics and drumlins　　　　　　　　　　Grade 4

2. Study **Figure 2**, a photo of an erratic boulder in North Yorkshire. Explain why erratics occur.　　**[2 marks]**

   ....

   ....

   ....

   ....

Figure 2

3. Complete the following sentences.　　**[2 marks]**

   Drumlins are long egg-shaped hills made from ....

   The axis of their elongated shape indicates the .... of the glacier.

 Made a start　 Feeling confident　 Exam ready

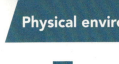

Economic activities

② Quick quiz

1. Complete the sentence by ticking **one** of the options below.

Arable farming is more suited to...

☐ ...upland areas.   ☐ ...lowland areas.

2. Give **one** negative impact of quarrying on glaciated landscapes.

....

⑤ Farming on glaciated landscapes — Grade 4

1. Study **Figure 1**, a photo of sheep farming in a glaciated upland landscape.

Figure 1

Suggest why upland glaciated landscapes are more suitable for pastoral farming. **[2 marks]**

> Pasture is land covered with grass and other low plants suitable for grazing animals, especially cattle or sheep.

> **Exam focus**
>
> In this 2-mark question, you will need to develop your point and say **why** the upland landscapes are more suitable.

> **Exam focus**
>
> Be sure to use the information in Figures in your answer. For example, as you can see from Figure 1, the slopes are very steep and/or the soil is rocky.

The soils in upland glaciated landscapes are rocky and thin, which means they lack the nutrients for crops to grow well.

....

2. Suggest **one** way farmers in glaciated upland landscapes can increase their income. **[1 mark]**

....

⑩ Opportunities and conflicts in glaciated landscapes — Grade 3

3. State **two** opportunities created by tourism in glaciated landscapes. **[2 marks]**

Opportunity 1 ....

Opportunity 2 ....

4. Give **one** reason why tourists can cause conflict with local residents in glaciated landscapes. **[1 mark]**

....

....

5. Outline **one** negative impact of forestry on glaciated landscapes. **[2 marks]**

....

....

 ✓ **Made a start**   ✓ **Feeling confident**   ✓ **Exam ready**

Tourism in Snowdonia

② Quick quiz

1. Complete the sentence.

Tourism can have a positive economic impact on upland areas, as it provides .... for local people.

2. Name **two** outdoor recreation opportunities offered by glaciated upland landscapes.

....

⑤ Tourism management strategies   Grade 4

1. Outline **one** strategy that can be used to manage the impact of tourism in glaciated upland areas. **[2 marks]**

One strategy that can be used to manage the impact of tourism
is to mark specific footpaths through environmentally sensitive
areas
....
....

> **Named example**
>
> For your exam, you need to know about a glaciated upland area in the UK used for tourism. This page uses Snowdonia National Park in Wales, but you should use the example you have studied. Make sure you know specific examples of benefits of, and conflicts resulting from tourism for your example.

⑩ Tourism in glaciated landscapes   Grade 3

2. Figure 1 shows a tourist attraction in Snowdonia National Park.

Figure 1

(a) Using **Figure 1**, give **one** reason why tourists visit glaciated upland landscapes. **[1 mark]**

....

(b) Suggest **one** positive environmental impact of tourism in glaciated upland landscapes. **[1 mark]**

....

3. Suggest **one** negative economic impact of tourism in glaciated upland landscapes. **[2 marks]**

....
....

 Made a start   **Feeling confident**   **Exam ready**

Levelled response questions

 Waterfall formation

1. Study **Figure 1**, a photo of a waterfall.
Explain how waterfalls change over time.

[6 marks]

Figure 1

Figure 1 shows a waterfall with a plunge pool beneath it.

....

....

....

....

....

....

....

....

....

....

....

....

Use the image to identify key features (e.g. plunge pool, harder, resistant rock). In your answer, use phrases such as '**Figure 1** shows a waterfall with …'

Exam focus

In 6-mark levelled response questions, you need to:
• provide detailed explanations
• accurately use specialist terminology
• make use of any additional resources provided.

Continue your answer on your own paper if you need to.

 Cliffs and wave-cut platforms

2. Study **Figure 2**, a photo of the Seven Sisters cliff in Sussex. Describe and explain how the landform shown changes over time. **[6 marks]**

Figure 2

....

....

....

....

....

....

....

....

....

....

....

You must structure your answer logically. Make clear links answer to this question the interaction of processes and the landforms shown in **Figure 2**.

Continue your answer on your own paper.

 Made a start  **Feeling confident** **Exam ready**

Urbanisation characteristics

BBC

---

⑤ Quick quiz

1. What is a megacity?

....

2. Give **one** cause of counter-urbanisation.

....

3. Is the statement below true or false? Circle the correct answer.

A push factor is a reason why people want to leave a place.

true   false

4. Complete the sentence.

.... migration, when people move from rural to .... areas, affects the .... of urbanisation.

---

⑮ Urban patterns and changes   **Grade 3**

1. Study **Figure 1**, which shows the percentage of people living in urban areas around the world.

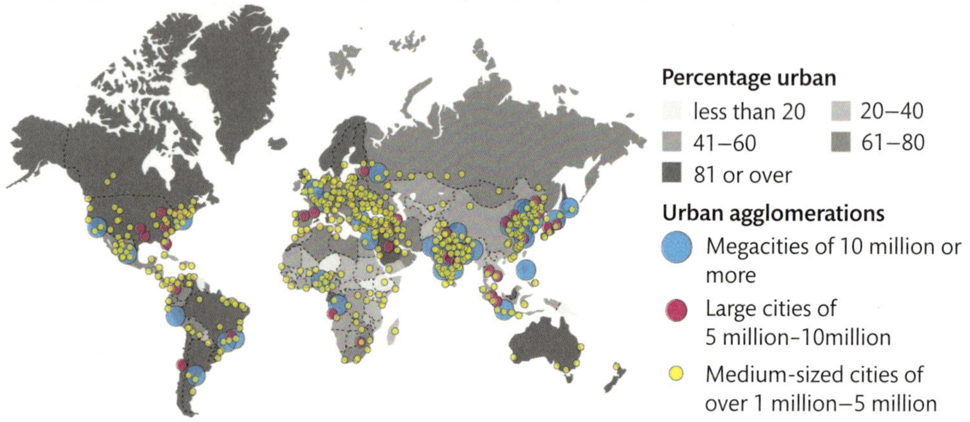

Percentage urban
- less than 20
- 41–60
- 81 or over
- 20–40
- 61–80

Urban agglomerations
- 🔵 Megacities of 10 million or more
- 🔴 Large cities of 5 million–10million
- 🟡 Medium-sized cities of over 1 million–5 million

Exam focus 📌
Remember to use evidence from any maps and graphs that are provided in your exam.

Compare the variations in Asia with North America.

Figure 1

Using **Figure 1**, describe the distribution of megacities.   **[3 marks]**

Figure 1 shows that the majority of megacities are located in the northern hemisphere, with more ....

megacities in ....

2. Explain **one** factor affecting the rate of urbanisation.   **[2 marks]**

One factor that affects the rate of urbanisation is natural increase ....

....

3. Complete the following sentences using some of the words below.   **[2 marks]**

| urban | reversed | the same | rural | megacities |

In 1950, two-thirds of the world's population lived in .... areas and one-third in .... areas. By 2050, this is likely to be ...., with more than 6 billion people living in urban areas.

4. Urbanisation is happening faster in lower income countries (LICs) and newly emerging economies (NEEs) than in higher income countries (HICs). Using **Figure 1** and your understanding, suggest reasons for the patterns shown.   **[4 marks]**

....

....

....

 Made a start   **Feeling confident**   ✓ **Exam ready**

Human environments / **Urban issues** / **Urbanisation**

Case study: Opportunities in LICs and NEEs

---

② **Quick quiz**

1. Complete the sentence. Choose the correct phrase from the options below.

| counter–urbanisation | rural–urban migration | GDP |

The two main reasons for rapid urban growth in cities are natural increase and ....

2. Name **one** social opportunity provided by urban growth. ....

3. What does the abbreviation GDP stand for? ....

---

⑤ **Causes of rapid growth** — **Grade 3**

1. Study **Figure 1**, a photo showing a woman farming in a rural area in India.

Figure 1

> **Case study**
>
> For your exam, you need to know a case study of a major city in an LIC or NEE. Revise the case study you have studied, including key facts such as the city's location and its importance nationally and internationally.

> Think about push and pull factors which contribute to rural-urban migration.

Suggest **one** reason why the kind of work shown in **Figure 1** might increase rural-urban migration in lower income countries (LICs) and newly emerging economies (NEEs). **[2 marks]**

In rural areas of LICs and NEEs, job opportunities are limited, and jobs like farming, as shown in ....

Figure 1, are ....

....

---

⑤ **Opportunities created by rapid economic growth** — **Grade 4**

2. Outline **one** economic opportunity created by rapid urban growth. **[2 marks]**

....

....

....

> **Named example**
>
> Refer to your case studies or named examples in your answers, even though the questions do not specifically ask you to.

3. Explain **one** social opportunity created by rapid urban growth. **[2 marks]**

....

....

☑ **Made a start**   ☑ **Feeling confident**   ☑ **Exam ready**   **53**

BBC

Case study: Challenges in LICs and NEEs

② Quick quiz

1. Identify the type of settlement shown in **Figure 1**.

....

2. State **one** challenge associated with the informal employment sector.

....

Figure 1

⑩ Challenges of rapid urban growth

Grades 3–9 ☑

1. Study **Figure 2**, a photo showing environmental pollution in a river in Mumbai.

Outline **one** environmental challenge of rapid urban growth in a named lower income country (LIC) or newly emerging economy (NEE) city.   **[2 marks]**

....

....

....

....

....

Figure 2

2. Explain how urban planning in a named LIC or NEE city is improving people's quality of life.   **[6 marks]**

....

....

....

....

....

....

....

....

....

....

....

Exam focus 📌

Revise all of the social, environmental and economic challenges created by rapid urban growth in your case study city.

Case study 🔍

For question **2**, you need to write about specific strategies used by your LIC or NEE case study city. Remember to include some specific facts from your case study to support the points you make.

☑ **Made a start**   ☑ **Feeling confident**   ☑ **Exam ready**

Distribution of UK population and cities

② Quick quiz

1. Complete the sentence.

In 2014, more than 80 per cent of people in the UK were living in .... areas.

2. Is the statement below true or false? Circle your answer.

Scotland has more cities than any other country in the UK.

true   false

⑮ UK population distribution   **Grade 3**

1. Study **Figure 1**, a map showing population density in the UK.

(a) What is the population density of Powys?   **[1 mark]**

....

(b) Identify **one** city in England with a high population density. **[1 mark]**

....

(c) Suggest **one** disadvantage of using choropleth maps.   **[1 mark]**

....

....

....

Glasgow

Edinburgh

Belfast

Manchester

Liverpool

Birmingham

Powys

Cardiff

London

14 500   9600   4800   2400   1200   600   0

people per square km

Figure 1

2. Using **Figure 1**, describe the distribution of areas of high population density in the UK.   **[3 marks]**

The area of the highest population density in the UK is ....

London, in the south-east of England. The north-west ....

....

....

....

For the choropleth map in **Figure 1**, the darker areas indicate a higher population density.

3. Suggest how physical and human factors affect population density.   **[4 marks]**

....

....

....

....

....

....

....

Case study: Opportunities in the UK

② Quick quiz

1. Complete the sentence.

An .... transport system is different transport methods connected together.

2. Is the statement below true or false? Circle the correct answer.

Being near to several large airports makes a city more attractive to businesses looking to locate there.

true false

⑤ City opportunities   Grade 3

1. Outline **one** social opportunity created by rapid urban growth in a named UK city.   **[2 marks]**

London has a huge variety of recreational opportunities, including ....

....

....

....

> **Case study**
>
> For your exam, you need to know a case study of a major UK city. Make sure you revise both the opportunities and challenges created by urban change in the case study city you have studied.

2. Outline **one** economic opportunity created by rapid urban growth in a named UK city.   **[2 marks]**

An economic opportunity created by urban growth in London is the huge range of employment ....

opportunities, ...

....

⑩ Impacts of migration   Grades 5–9

3. Migration has created many benefits for UK cities. Do you agree with this statement? Use your own understanding to support your answer.   **[6 marks]**

....

....

....

....

....

....

....

....

....

> **Exam focus**
>
> For a named UK city, make sure you describe and explain several different impacts, both positive and negative. You will need to make an informed judgement by weighing up the positive and negative impacts, and link your ideas back to the question to give evidence of how these have affected your chosen city. For example, use phrases like 'because', 'this has contributed to' or 'this has led to'.

> The model answer refers to London – use the named example you have studied to answer this question.

> Continue your answer on your own paper.

✓ **Made a start**   ✓ **Feeling confident**   ✓ **Exam ready**

Case study: Challenges in the UK

---

② **Quick quiz**

1. Complete the sentence.

   Rises in house .... in cities

   have increased the number of ....

   settlements.

2. Name **one** negative environmental impact of urban sprawl.

   ....

3. Name **one** challenge of building on brownfield sites.

   ....

4. Name **one** challenge related to education caused by urban growth.

   ....

---

⑤ **Urban regeneration** **Grade 3**

1. Outline **one** reason for an urban regeneration project in a named UK city. **[2 marks]**

   ....

   ....

   ....

   ....

> **Case study**
>
> For your exam, you need to know about an example of an urban regeneration project in a UK city. The answers for this page use the example of the Lower Lea Valley Redevelopment Project in London, but you should revise the example you have studied.

2. Give **one** benefit of an urban regeneration project you have studied. **[1 mark]**

   ....

   ....

> Think about the key problems that an area of deprivation faces, and why councils and local governments would want to make improvements.

---

⑩ **Challenges of rapid economic growth** **Grades 5–9**

3. Urban change has created economic challenges in UK cities.
   Do you agree with this?
   Using an example you have studied, explain your answer. **[6 marks]**

   ....

   ....

   ....

   ....

   ....

   ....

   ....

   ....

   ....

> Continue your answer on your own paper.

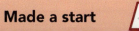 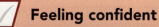

Urban sustainability

---

⑤ Quick quiz

1. Study **Figure 1**. Identify the feature of sustainable urban living shown.

   ....

2. Name a water-stressed city.

   ....

3. Name **one** strategy for conserving water.

   ....

Figure 1

---

⑩ Sustainable urban living   Grades 3–4

1. Study **Figure 2**, a photo of solar panels on the roofs of houses.

   Explain **one** way in which solar panels can contribute to sustainable urban living.   **[2 marks]**

   🚩 Energy conservation is an important part of sustainable ....

   urban living. ....

   ....

   ....

   ....

Figure 2

2. Explain **one** benefit of creating green spaces in urban environments.   **[2 marks]**

   ....

   ....

   ....

3. Outline **one** sustainable urban transport scheme.   **[2 marks]**

   ....

   ....

   ....

   ....

   ....

> **Exam focus** 📌
>
> Remember, in an **outline** question there is no need to give an explanation. You just need to set out the characteristics of the thing you're being asked about.

4. Describe **one** sustainable urban waste recycling scheme.   **[2 marks]**

   ....

   ....

   ....

 Made a start   **Feeling confident**   **Exam ready**

Classifying development: economic measures

② Quick quiz

1. What does LIC stand for?

....

2. What does NEE stand for?

....

② Economic measures and their limitations

Grade 3

1. Outline **one** economic measure of development. **[2 marks]**

> One economic measure of development is gross national
>
> income. GNI is ....
>
> ....
>
> ....

This is an **outline** question, so you need to name the measure and say what it is.

2. Explain **two** limitations of economic measures of development. **[4 marks]**

....

....

....

....

⑤ Global GNI per capita

Grade 3

3. Study **Figure 1**, a map showing worldwide gross national income (GNI) per capita by country in 2014.

thousands (current US$)

- <6.80
- 6.80–16.23
- 16.23–29.29
- 29.29–51.75
- >51.75

Figure 1

Geographical skills

Although choropleth maps are useful for providing a visual representation of data, you also need to know their limitations. Here, the map shows how GNI differs between countries, but it does not represent the variations in wealth within a country.

(a) Identify a country with a GNI per capita of less than US$6800. **[1 mark]**

....

(b) Identify a high income country. **[1 mark]**

....

(c) Identify a country with a GNI per capita of US$29 290–US$51 750. **[1 mark]**

....

Classifying development: social measures

② Quick quiz

1. How are birth and death rates measured?

☐ per 100 people

☐ per 1000 people

2. Complete the sentence.

Infant mortality is the number of ....
who die before their .... birthday.

3. Name the **two** measures of development that are used, alongside GNI per capita, to calculate a country's HDI score.

....

⑤ Development indicators   Grade 3

1. Give **one** reason for high infant mortality rates in a country.   **[1 mark]**

....

> Think of the types of services that people in less developed countries may not have access to.

2. Explain **one** advantage and **one** limitation of social measures of development.   **[4 marks]**

An advantage of social measures of development is that they ....

give a reliable indication of the standard of living in a country. For ....

instance, low death rates suggest that the healthcare system is ......

very good. A limitation of social measures of development is ....

....

....

....

Exam focus

Make sure you understand the connection between what a development measure shows, and what it suggests about a country's development. For example, a low birth rate may indicate a country has established a good education system, as higher levels of education contribute to more women choosing a career before, or instead of, having a family. This suggests the country is likely to be more developed.

⑤ Variations in HDI across Africa   Grade 3

3. Study **Figure 1**, a map showing the Human Development Index (HDI) score of countries in Africa. Identify the HDI for Chad.   **[1 mark]**

....

4. Describe the pattern of HDI scores in Africa.   **[2 marks]**

....

....

5. Suggest **one** advantage of using HDI as a measure of a country's level of development.   **[2 marks]**

....

....

Human development index (1 = high/ 0 = low)

■ <0.42   ■ 0.42–0.47   ■ 0.48–0.51
■ 0.52–0.56   ■ 0.57–0.68   ■ >0.68   ■ no data

Figure 1

 Made a start   **Feeling confident**   ☑ **Exam ready**

The Demographic Transition Model

② Quick quiz

1. Which stage of the Demographic Transition Model (DTM) does the statement below describe?
Tick the correct answer.

Birth and death rates are both high and fluctuating.

☐ Stage 5   ☐ Stage 1

2. Which stage of the DTM does the statement below describe?
Tick the correct answer.

Birth and death rates are both low and fluctuating.

☐ Stage 2   ☐ Stage 4

3. Draw lines to match each society to the stage of the DTM you think it is at.

Mexico	Stage 1
Indigenous rainforest tribes	Stage 4
the USA	Stage 3

⑤ Understanding the DTM   Grade 3

1. Suggest what a declining birth rate may indicate about a country's development. **[2 marks]**

🚩 Declining birth rates suggest that a country has improved

access to ....

....

> Think about the stage of the Demographic Transition Model where birth rates begin to decline.

2. Explain **one** reason why the total population starts to decline in Stage 5 of the DTM. **[2 marks]**

🚩 The total population starts to decline in Stage 5 of the

DTM because ....

....

....

Geographical skills

In your exam, you may be asked to label, complete or draw conclusions from a DTM. Make sure you know what is happening to the birth rate, death rate and overall population size for each stage of the model.

⑤ Stage 2 of the DTM   Grade 4

3. Explain the reasons why death rates fall in Stage 2 of the DTM. **[4 marks]**

....

....

....

....

....

....

....

Exam focus

The command word **explain** means you need to give reasons. For a 4-mark question, you could give two developed reasons or one reason that is clearly developed with three points.

Uneven development: causes and consequences

② Quick quiz

1. Which causes of uneven development are shown in these pictures – physical, economic or historical?

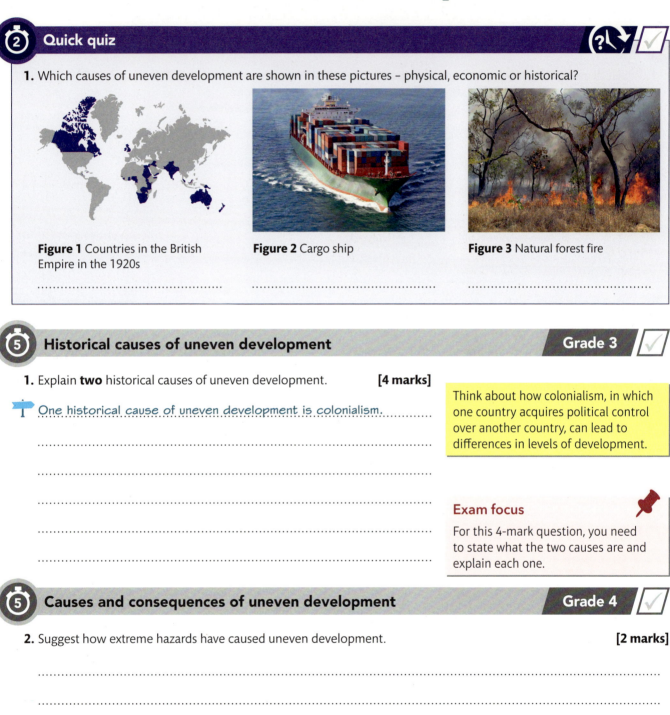

Figure 1 Countries in the British Empire in the 1920s

....

Figure 2 Cargo ship

....

Figure 3 Natural forest fire

....

⑤ Historical causes of uneven development   Grade 3

1. Explain **two** historical causes of uneven development.   **[4 marks]**

🚏 One historical cause of uneven development is colonialism. ....

....

....

....

....

....

> Think about how colonialism, in which one country acquires political control over another country, can lead to differences in levels of development.

> **Exam focus** 📌
>
> For this 4-mark question, you need to state what the two causes are and explain each one.

⑤ Causes and consequences of uneven development   Grade 4

2. Suggest how extreme hazards have caused uneven development.   **[2 marks]**

....

....

....

....

3. Explain **one** consequence of uneven development.   **[2 marks]**

....

....

....

....

> State the consequence, and explain briefly how it is caused by uneven development.

✓ **Made a start**   ✓ **Feeling confident**   ✓ **Exam ready**

Reducing the global development gap

② Quick quiz

1. Complete the sentence by ticking **one** of the options below.

Food and medical supplies donated from one country to another are an example of...

☐ ...fairtrade.

☐ ...short-term aid.

2. Complete the sentence using some of the words below.

Writing off the .... of poorer countries enables them to .... in improving the population's quality of life.

| invest | healthcare | debt |

---

⑤ Investment, intermediate technology and tourism [Grade 2] ☐

1. State **one** aim of intermediate technology? **[1 mark]**

🪧 Intermediate technology aims to meet the needs of ....

....

2. Explain **one** way in which investment can help to reduce the development gap. **[2 marks]**

....

....

3. Suggest how tourism can help reduce the development gap for a place you have studied. **[2 marks]**

....

....

....

Exam focus 📌

For a 2-mark **explain** question, you should briefly develop your point. In question **2**, you need to think about **how** investment affects development.

Named example 📍

Include details from the examples you have studied. You could mention a particular tourism project that has reduced the development gap, or the impacts of tourism in your chosen LIC or newly emerging economy (NEE) more generally.

---

⑤ Reducing the development gap [Grade 4] ☐

4. Suggest how fairtrade can help reduce the development gap. **[2 marks]**

....

....

....

5. Outline **one** way in which financial investment from transnational corporations (TNCs) or higher income countries (HICs) can help reduce the development gap in LICs. **[2 marks]**

....

....

....

....

 Made a start ☑ Feeling confident Exam ready

Case study: Developing LICs and NEEs

② Quick quiz

1. Decide whether the following examples are more likely to increase or reduce development. Tick **one** choice for each.

		Increase	Reduce
(a)	growing film industry		
(b)	seasonal monsoons		
(c)	well-connected sea ports		
(d)	tourism		

⑤ Trading relationships and TNCs   Grade 4

1. For a lower income country (LIC) or newly emerging economy (NEE) you have studied, outline **one** way trading relationships are contributing to rapid economic development.   **[2 marks]**

> Stronger trading relationships often improve the value of imports and exports.

....

....

....

2. For a lower income country (LIC) or newly emerging economy (NEE) you have studied, outline **one** way transnational corporations (TNCs) are contributing to rapid economic development.   **[2 marks]**

> **Case study**
>
> You need to know a case study of one LIC or NEE which is experiencing rapid economic development. You should revise the country you have studied, and learn key facts about your country such as its location and its cultural, political and environmental context.

....

....

....

⑩ Social and environmental context   Grade 4

3. For an LIC or NEE you have studied, state a social factor that has contributed to the development of the country.   **[1 mark]**

....

....

....

4. For an LIC or NEE you have studied, explain how environmental factors have affected its development.   **[4 marks]**

....

....

....

....

 Made a start   **Feeling confident**   **Exam ready**

Case study: Impacts of development

Quick quiz

1. Give **one** environmental impact of rapid economic development.

....

2. Name **two** types of international aid that may contribute to the development of a newly emerging economy (NEE) or lower income country (LIC).

....

Investment, intermediate technology and tourism — Grade 4

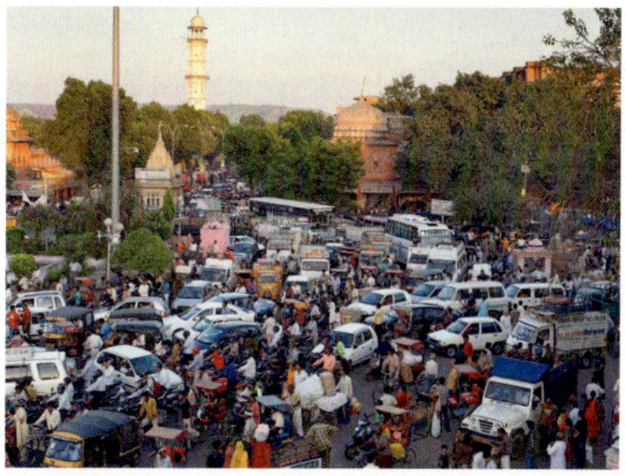

Figure 1

1. Study **Figure 1**, a photo showing an urban area experiencing rapid economic development. Suggest **one** impact of rapid economic growth in cities in NEEs. **[2 marks]**

....

....

2. For an LIC or NEE you have studied, state **two** advantages and **two** disadvantages of foreign direct investment from transnational corporations (TNCs). **[4 marks]**

Advantages Foreign investment has helped to improve infrastructure. ....

....

Disadvantages ....

....

Impacts of rapid economic development — Grades 5–9

3. For an LIC or NEE you have studied, to what extent are the impacts of rapid urban development in urban areas mainly positive? **[6 marks]**

Exam focus

For a 6-mark question your answer needs to have a logical structure, accurately use specialist terminology, and provide detailed explanations that demonstrate your knowledge and understanding.

Continue your answer on your own paper.

....

....

....

....

....

....

....

....

....

....

Made a start | Feeling confident | Exam ready

Causes of economic change

② Quick quiz

1. Give an example of a primary sector job. ....

2. Give an example of a tertiary sector job. ....

3. Complete the sentence by circling the correct words below.

   Globalisation is the **increased / decreased** movement of people and goods between **jobs / countries**.

⑩ The changing UK economy   Grade 3

1. Explain the reasons for deindustrialisation in the UK.   **[4 marks]**

   One reason for deindustrialisation in the UK is globalisation, which has

   made it cheaper to import many products than to produce them in the UK. A second reason for

   deindustrialisation in the UK

   ....

   ....

> Think about how developments in technology have brought about economic change.

2. Outline **one** way deindustrialisation has affected employment in different regions of the UK.   **[2 marks]**

   ....

⑩ UK economic sectors   Grade 4

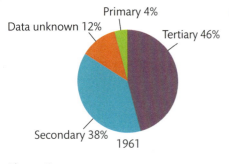

Primary 4%
Data unknown 12%
Tertiary 46%
Secondary 38%
1961

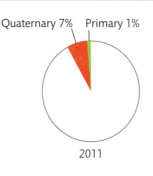

Quaternary 7%   Primary 1%
2011

Geographical skills

To draw a pie chart, you need to represent each part of the data as a proportion of 360 (the number of degrees in a circle). For example 12% of 360° is $\frac{12}{100} \times 360°$, or 43°.

Figure 1

3. Study **Figure 1**, two pie charts showing the UK's changing employment structure in 1961 and 2011.

   (a) What was the employment share for the secondary sector in 1961?   ....   **[1 mark]**

   (b) The table below shows the percentage of employment share for the secondary and tertiary sectors in 2011. Complete **Figure 1** with this data.   **[2 marks]**

Sector 2011	Tertiary	Secondary	Quaternary	Primary
employment share	80%	12%	7%	1%

   (c) Calculate the increase of the tertiary sector percentage employment share from 1961 to 2011.

   ....   **[1 mark]**

4. Outline **one** reason why the employment share for the quaternary sector has increased.   **[2 marks]**

   ....

 Made a start   **Feeling confident**   **Exam ready**

Impacts of industry

② Quick quiz

1. What is the impact shown in **Figure 1**?

2. What is the impact shown in **Figure 2**?

....

....

Figure 1

Figure 2

---

② Sustainable industry   Grade 3 ✓

1. Outline **one** way modern industries can operate more sustainably.

[2 marks]

> Think about why companies might adopt more sustainable strategies and the associated benefits.

✸ Industries can operate more sustainably by limiting their ....

environmental impacts. For example, ...

....

....

---

⑤ Environmental impacts   Grade 5 ✓

2. Explain the impact of industrial operations on the UK environment.

[4 marks]

....

....

....

....

....

....

....

....

....

....

....

....

Named example

You need to know an example of the environmental consequences of modern industrial development and ways in which organisations can be more environmentally sustainable.

Exam focus

For 4-mark explain questions, you can either make two developed points or one clear point with three further points that link and develop, explain and exemplify.

Clear description – Chemical industries can produce waste products that cause environmental pollution.

Explanation – Waste gases, such as sulfur dioxide, can be emitted into the atmosphere and attach to water vapour causing acid rain. Acid rain can increase soil acidity, which can cause woodland and aquatic life to decline.

Rural landscape changes

(2) Quick quiz

1. State **one** rural area experiencing population growth.

2. State **one** rural area experiencing population decline.

....

....

(10) Rural growth and decline **Grade 3**

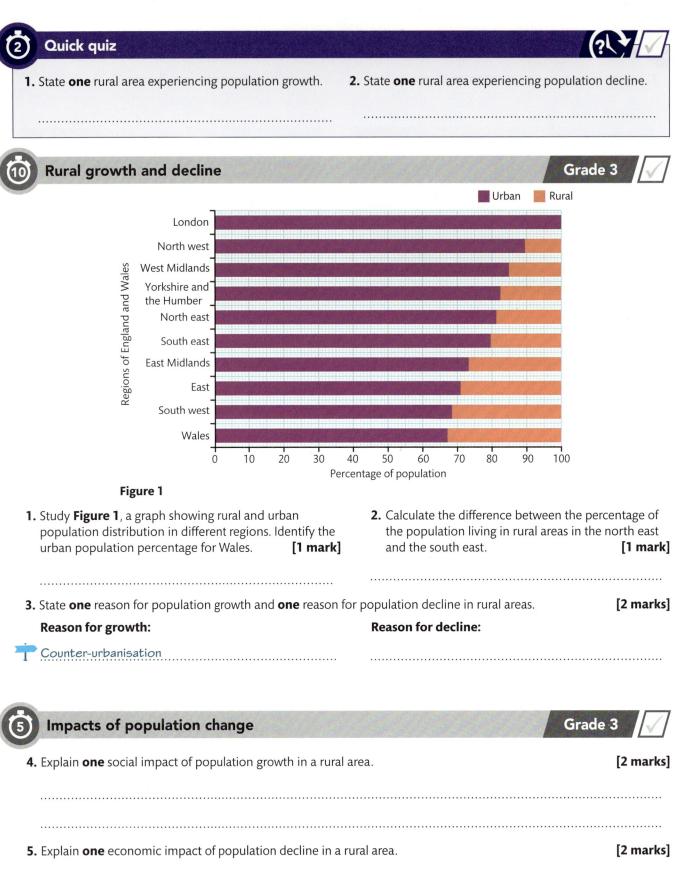

Urban · Rural

Figure 1

1. Study **Figure 1**, a graph showing rural and urban population distribution in different regions. Identify the urban population percentage for Wales. **[1 mark]**

....

2. Calculate the difference between the percentage of the population living in rural areas in the north east and the south east. **[1 mark]**

....

3. State **one** reason for population growth and **one** reason for population decline in rural areas. **[2 marks]**

Reason for growth:

Counter-urbanisation.................................

Reason for decline:

....

(5) Impacts of population change **Grade 3**

4. Explain **one** social impact of population growth in a rural area. **[2 marks]**

....

....

5. Explain **one** economic impact of population decline in a rural area. **[2 marks]**

....

....

 Made a start **Feeling confident** ☑ **Exam ready**

Infrastructure improvements

N E

② Quick quiz

1. Complete the sentence using some of the words below.

In the UK, ... increase is putting pressure on .... infrastructure.

| transport | land | population | geographic |

2. Name **one** new UK rail development scheme.

....

⑤ UK airport and railway developments — Grade 3

1. Outline **two** benefits of improvements to airports in the UK.   **[2 marks]**

> Think about the wider benefits if more people used UK airports.

🚩 One benefit is increased employment opportunities and a ....

second is ....

....

2. Outline **one** reason for a UK railway improvement scheme.   **[2 marks]**

....

....

⑤ UK ports — Grade 3

3. Study **Figure 1**, a graph showing the total imports and exports handled by UK ports between 1980 and 2015.

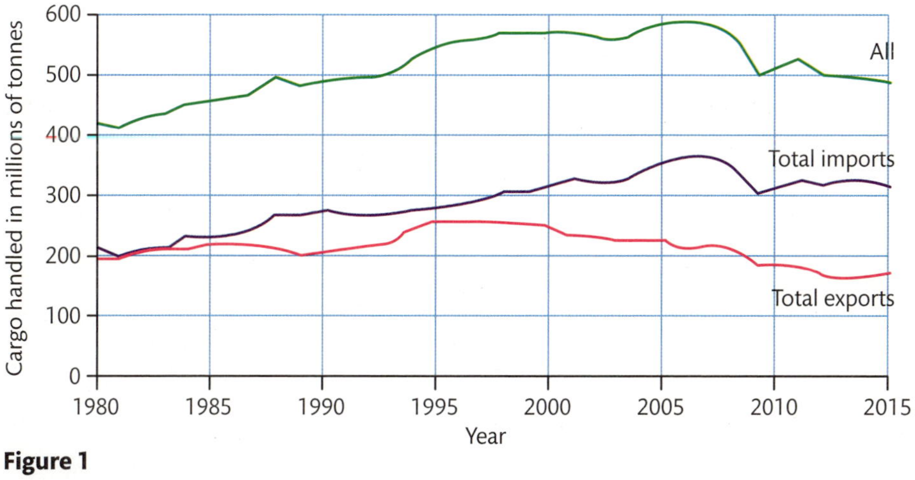

Figure 1

Graph: Cargo handled in millions of tonnes (y-axis, 0–600) vs Year (x-axis, 1980–2015). Lines labelled All, Total imports, Total exports.

> **Exam focus** 📌
> Take your time when answering questions about a graph in your exam. Use a ruler to make sure you are reading data accurately from the graph.

(a) What was the total tonnage of exports handled by UK ports in 1980?   ....   **[1 mark]**

(b) What was the total tonnage of imports handled by UK ports in 2000?   ....   **[1 mark]**

(c) Using **Figure 1**, suggest how improvements to infrastructure may have affected imports handled by UK ports between 1980 and 2015.   **[2 marks]**

....

....

The north–south divide

BBC

---

② Quick quiz

1. Is the statement below true or false? Circle the correct answer.

Government spending on infrastructure is higher in London than anywhere else in the UK.

true   false

2. Complete the following sentence.

The north–south divide refers to

.... differences between the

two regions.

---

⑩ Evidence of and reasons for the north–south divide   **Grade 4**

1. Study **Figure 1**, which shows weekly earnings in the UK by region.

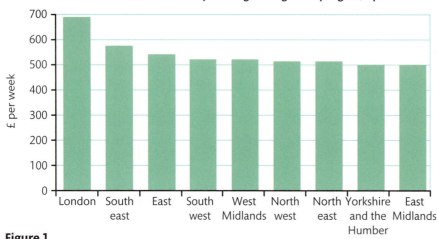

Median full-time weekly earnings in England by region, April 2017

Figure 1

Exam focus

There is more than one reason for the north–south divide. Make sure you learn all of them – you might be asked to provide more than one.

Identify the difference between weekly earnings for Yorkshire and the Humber and London.   **[1 mark]**

....

2. Suggest **one** reason why the north-south divide exists in the UK.   **[2 marks]**

One of the reasons for the north–south divide in the UK is the decline of primary industry in ....

....

3. Figure 1 shows the difference in average earnings in the north and the south. Suggest **one** other measure that could be used to determine the extent of the north–south divide.   **[1 mark]**

....

---

⑤ Bridging the north–south divide   **Grade 4**

4. Explain the strategies for resolving the regional differences in the UK.   **[4 marks]**

....

....

....

....

 Made a start   **Feeling confident**   **Exam ready**

The UK in the wider world

② Quick quiz

1. Complete the sentence.

The Commonwealth is a ....

organisation of 52 ....

3. Is the statement below true or false? Circle the correct answer.

Around 375 million people around the world speak English as their first language.

true   false

2. Name **one** advantage of EU membership for the UK.

....

4. Complete the sentence.

A superfast .... network across the UK has helped businesses to communicate globally and .... their operations worldwide.

⑮ Benefits of links with the wider world   Grade 4

1. Outline **one** of the UK's transport links with the wider world.   **[2 marks]**

....

2. Give **two** benefits of the UK's links with the wider world.   **[2 marks]**

Benefit 1 Increased tourism provides more employment .... opportunities. ....

Benefit 2 ....

....

> **Exam focus**
>
> This is a **give** question, so you can state each benefit without providing any explanation.

3. Outline **one** benefit of the Commonwealth.   **[2 marks]**

Trade between Commonwealth countries is ....

....

....

4. Study **Figure 1**, a graph showing the UK's top trading partners in 2015 as a percentage of total UK trade.

(a) Complete the graph to show the value of the UK's exports to the USA in 2015 (47 000 million pounds). **[1 mark]**

(b) Which country did the UK export approximately £18 million of goods to in 2015? **[1 mark]**

....

(c) Calculate the approximate difference between exports to Switzerland and Germany in millions of pounds. **[1 mark]**

....

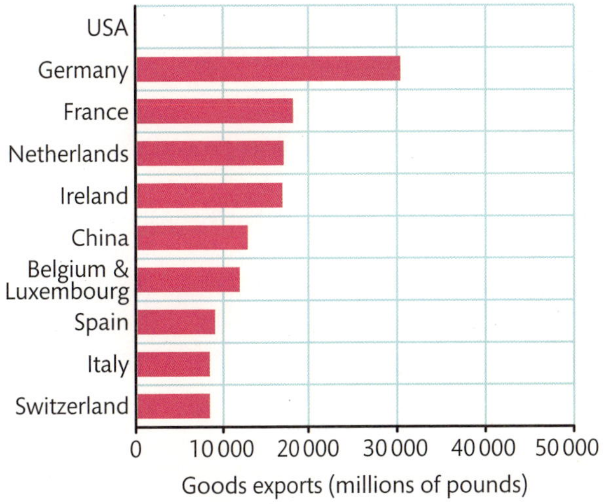

Figure 1: bar chart — Goods exports (millions of pounds) by country: USA, Germany, France, Netherlands, Ireland, China, Belgium & Luxembourg, Spain, Italy, Switzerland. X-axis 0, 10 000, 20 000, 30 000, 40 000, 50 000.

Figure 1

Types of resources

② Quick quiz

1. Complete the sentence.

.... is needed to power factories,

transport, homes, schools and businesses.

3. What are the **three** most important resources for human development?

....

2. The Middle East is one of the largest producers of which energy resource?

....

4. State **two** things fresh water is needed for.

....

⑤ Global calorie consumption   Grade 2

1. Study **Figure 1**, a map showing the average daily calorie consumption per capita of countries around the world.

    (a) What is the daily calorie intake per capita for Brazil?
    [1 mark]

    ....

    ....

    (b) Suggest **one** reason why countries like the USA have a high daily calorie intake per capita.   **[2 marks]**

    ....

    ....

    ....

    ....

USA

Brazil

- ■ over 3000
- ■ >2500 – ≤3000
- ■ ≥2000 – ≤2500
- ■ less than 2000

Figure 1

Geographical skills

A choropleth map represents data using different shades of colours. When you are obtaining data make sure you double check your answer, as maps using several similar colours can sometimes be difficult to read against the key.

⑤ Global inequalities in resources   Grade 4

2. Outline **one** reason for global inequalities in water supply.   **[2 marks]**

....

....

....

3. Outline **one** reason for global inequalities in energy supply.   **[2 marks]**

....

....

....

☐ **Made a start**   ☐ **Feeling confident**   ☐ **Exam ready**

Food resources in the UK

② Quick quiz

1. Give **one** reason why demand for food in the UK has increased.

....

3. Complete the sentence.

There is a demand for ....food types

at affordable .... in the UK.

2. Complete the sentence by ticking **one** of the options below.

The sales of organic food in the UK have...

☐ ...increased.   ☐ ...decreased.

4. Name **one** high value food import.

....

⑤ Food miles — Grade 3

1. Study **Figure 1**, an infographic showing examples of food miles for different foods. Suggest **one** reason for the increase in food miles.

[2 marks]

One reason for increasing food miles is ....

....

....

2. Outline the impact of increasing food miles. [2 marks]

Importing food products from countries further away increases

food miles, ...

....

conventional	local
2000 miles	50 miles
2000 miles	40 miles
3000 miles	20 miles
6000 miles	10 miles

Figure 1

② UK supply and demand — Grade 5

3. Which **one** of the following statements about UK food consumption is true? Shade **one** circle only. [1 mark]

A The UK is self-sufficient in vegetables, but not in fruit. ○

B Everyone eating more is the main reason for increasing demand for food in the UK. ○

C There is a demand for some products all year round which cannot be grown in the UK. ○

D Buying locally grown, seasonal produce does not affect the amount of food the UK imports. ○

⑤ Agribusiness — Grades 4–5

4. Explain how agribusinesses can help to meet the rising demand for food in the UK. [4 marks]

....

....

....

....

☑ Made a start   ☑ Feeling confident   ☑ Exam ready

Water resources in the UK

② Quick quiz

1. What are the **three** main sources of the UK's water?

....

3. Complete the sentence.

Water .... schemes maintain

water supplies by moving water from areas of

.... to areas of deficit.

2. Complete the sentence by ticking **one** of the options below.

The north and west of the UK experience a water...

☐ ...surplus.   ☐ ...deficit.

4. Complete the sentence.

Average water use per .... per day

has been .... since 1930.

⑤ Water stress and water quality   Grade 3

1. Outline the reasons for water stress in the south and east of the UK.
[2 marks]

> Think about the balance between precipitation and population.

🪧 In the south and east of the UK, the demand for water is high due

to high population density, and lower rates of ....

....

2. Explain how water quality is monitored in the UK.   **[2 marks]**

🪧 The Environment Agency constantly checks water quality for ....

pollution and ....

....

Geographical skills 🌐

A distribution refers to the way geographical features are spread out or arranged. When describing distribution on a map, as in question **3**, use PQE:
P – Give the general **pattern**.
Q – **Quantify** (support) the pattern.
E – Identify any **exceptions** (anomalies).

⑤ Rainfall distribution in the UK   Grade 3

3. Study **Figure 1**, a map showing the amount of rainfall across the UK in the summer.
Describe the distribution of rainfall.   **[2 marks]**

....

....

....

....

....

4. Identify **one** region that receives 400–449 mm of rainfall.   **[1 mark]**

....

5. Identify **one** region that receives 350–399 mm of rainfall.   **[1 mark]**

....

Northern Scotland · Eastern Scotland · Western Scotland · Eastern England · Central England · East Anglia · North Wales and north west England · Southern England · South west England and South Wales

Rainfall (mm)
▢ 200–249 ▢ 250–299 ▢ 300–349
▢ 350–399 ▢ 400–449

Figure 1

 Made a start   **Feeling confident**   ☑ **Exam ready**

Energy resources in the UK

② Quick quiz

1. Complete the sentence.

   In the UK, ... for natural gas exceeds domestic supply.

2. How has the UK's reliance on coal and oil changed since 1970?

   ....

3. Name **two** sources of renewable energy which are increasingly being used instead of fossil fuels in the UK.

   ....

4. Give **one** reason UK coal mines have closed.

   ....

⑤ Issues of exploiting energy   Grade 4

1. Explain how exploiting energy sources might cause economic and environmental issues. **[4 marks]**

   Economic issues associated with exploiting energy sources ....

   include the cost of building nuclear power plants, ...

   ....

   ....

   Environmental issues that may be caused by exploiting nuclear and

   fossil fuel energy sources include ....

   ....

> Make sure you revise the issues associated with all renewable and non-renewable energy sources, as you could be asked about any of them in your exam.

> **Exam focus**
>
> In a 4-mark **explain** question, include your geographical knowledge and understanding of the issues involved. Try to make relevant links between different topics.

⑤ Wind operating capacity   Grade 3

2. Study **Figure 1**, a graph showing the UK operating capacity of wind farms between 2000 and 2015.

   (a) Describe the changes in wind power operating capacity from 2000 to 2015. **[2 marks]**

   ....

   ....

   ....

   ....

   (b) Identify the overall operating capacity during 2006–2007. **[1 mark]**

   ....

   (c) Identify the overall operating capacity during 2013–2014. **[1 mark]**

   ....

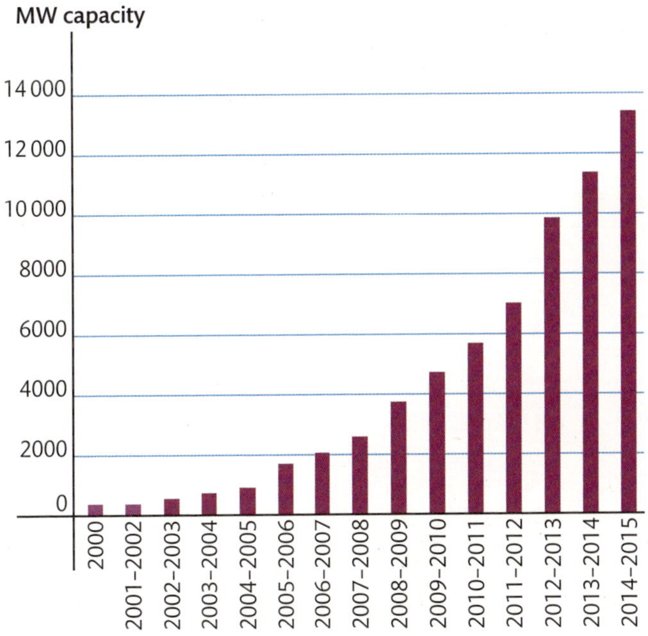

MW capacity

[Bar graph showing MW capacity on y-axis (0 to 14 000) and years on x-axis from 2000 to 2014–2015, with bars increasing over time]

Figure 1

Distribution of food

---

② Quick quiz

1. Complete the sentence.

Food .... means being without reliable access to enough affordable, nutritious food.

2. State **one** reason the global demand for food is increasing.

....

3. Give **one** reason food supply tends to be secure in higher income countries (HICs).

....

4. Complete the sentence.

Food consumption isn't equal. There is a .... surplus in many HICs and a deficit in many ....

---

⑤ Food insecurity   Grade 3

1. Explain **two** factors affecting food supply.   **[2 marks]**

🪧 Climate variations can affect crop yields. Rising temperatures...........

....

....

....

> There are a number of factors affecting food supply. Some are physical, such as variations in climate, others are economic, such as the amount of money countries can afford to invest in new technologies, and others are social, such as the effect of conflicts. Make sure you revise them all.

2. State **one** problem resulting from uneven global food consumption.   **[1 mark]**

....

---

⑤ Food consumption   Grade 2

3. Study **Figure 1**, a map showing global food production values for 2006.

(a) Identify a country whose food production value was more than $100 billion.   **[1 mark]**

....

(b) Using **Figure 1**, which **one** of the following statements about food production is true? Shade **one** circle only.   **[1 mark]**

A Food production values were higher in LICs than HICs.   ◯

B Most South American countries produced less than $1 billion worth of food.   ◯

C The UK produced more than $50 billion worth of food.   ◯

D Africa was the continent with the lowest food production value.   ◯

Legend:
- ■ >$100 billion
- ■ $50–100 billion
- ■ $20–50 billion
- ■ $10–20 billion
- ■ $5–10 billion
- ■ $1–5 billion
- ■ <$1 billion

Figure 1

 Made a start   **Feeling confident**   **Exam ready**

Impacts of food insecurity

② Quick quiz

1. Complete the sentence using the words below.

Food insecurity is when a .... does not have access to enough safe and nutritious food to have

an active life. It is the opposite of food ....

| supplier | security | safety | regulation | population |

2. Is the statement below true or false? Circle the correct answer.

More than 800 million people are affected by food insecurity around the world.

true   false

② Undernourishment   Grade 3

1. Which **one** of the following statements describes undernutrition?
Shade **one** circle only.   **[1 mark]**

A Undernutrition is when people consume less than 2000 calories a day. ◯

B Undernutrition is caused by eating lots of fatty, sugary food. ◯

C Undernutrition occurs when people do not have enough nutritious
food for growth, energy and a healthy immune system over a long
period of time. ◯

D Undernutrition occurs when people do not eat the recommended
amount of fruit and vegetables. ◯

> **Exam focus** 📌
>
> Multiple-choice questions can seem straightforward, but you should still read them carefully and make sure you've shaded the correct number of circles.

⑩ Impacts of food insecurity   Grade 4

2. Explain how food insecurity can cause soil erosion.   **[3 marks]**

....

....

....

....

....

3. (a) Outline how food insecurity can cause rising food prices.   **[2 marks]**

....

....

....

(b) Suggest **one** possible impact of a sudden rise in food prices.   **[1 mark]**

....

Increasing food supply

BBC

---

② Quick quiz

1. Complete the sentence using some of the words below.

The new .... revolution refers to the use of new technology to produce food, such as soil

conservation, developing varieties of seeds for particular growing .... and rainwater harvesting.

| brown | advantages | green | conditions | weather |

---

⑩ Large-scale agriculture **Grade 4**

1. Study **Figure 1**, a photo of plants growing in an aeroponic food farm.

Suggest **one** other strategy for increasing food supply. **[1 mark]**

....

2. Explain **two** advantages of a large-scale agricultural development in a named location. **[4 marks]**

📍 **Advantage 1** The Indus Basin Irrigation System in Pakistan has

helped to improve crop yields by giving farmers better access to

water to irrigate their crops.

Advantage 2 ....

....

....

....

3. Outline **one** disadvantage of a large-scale agricultural development.

....

....

Figure 1

Named example 📍

For your exam, you need to know an example of a large-scale agricultural development and its advantages and disadvantages. Think about the benefits for local people, the impact on the country's economy, and any environmental impacts.

[2 marks]

---

⑩ Managing food supply **Grades 5–9**

4. Explain how food supplies can be increased. **[6 marks]**

....

....

....

....

....

....

....

....

Exam focus 📌

For a 6-mark **explain** question, you need to show a thorough understanding of the interrelationships between people, processes and the environment.

Show clear cause and effect between actions and how they can provide a secure source of food.

Complete your answer on your own paper.

78

 Made a start **Feeling confident** **Exam ready**

Sustainable food options

② Quick quiz

1. Is the statement below true or false?
Circle the correct answer.

Organic farming is more costly than intensive agriculture, but also more sustainable.

true   false

2. Complete the sentence.

Setting .... helps to prevent overfishing.

3. Complete the sentence.

Biofertilisers, holistic grazing and agroforestry are all examples of ....

4. What is a zero-waste society?

....
....

⑤ Seasonal food consumption and urban farming   Grade 4

1. Outline how seasonal food consumption is sustainable.   **[2 marks]**

🚩 If more people buy locally produced, seasonal food, it reduces ....

....

....

> Think about how we source food products that can only be grown at specific times of the year in the UK.

> **Exam focus** 📌
> For a 2-mark **outline** question, you should give a brief overview in your answer, but you do not need to include lots of detail.

2. Explain how urban farming initiatives can make food supplies more sustainable.   **[2 marks]**

🚩 Urban farming initiatives can improve food supply sustainability ....

....

....

> **Named example** 📍
> For your exam, you need to know a named example of a local scheme in a lower income country (LIC) or newly emerging economy (NEE) to increase food supplies. You could use your named example in question **2**.

⑤ Permaculture and organic farming   Grade 4

3. Explain how permaculture is a sustainable food supply option.   **[2 marks]**

....

....

....

4. Explain how organic farming is a sustainable food supply option.   **[2 marks]**

....

....

....

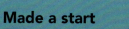 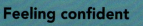

Distribution of water

② Quick quiz

1. Complete the sentence.

Water .... can be defined as the ability to access sufficient quantities of good quality fresh water for health, livelihood and production.

2. State **two** reasons why global water consumption is increasing.

....

....

⑩ Water insecurity   Grade 3

1. Outline **two** factors that affect water availability.   **[2 marks]**

> Think about a factor linked to wealth.

🚩 **Factor 1**   Climate variations mean that some countries may have a water surplus, while others have a water deficit. For example, a hot climate with low annual rainfall will result in a water deficit.

Factor 2 ....

....

2. Explain how over-abstraction can lead to water insecurity.   **[2 marks]**

....

....

3. Describe the global distribution of areas with a water deficit.   **[3 marks]**

🚩 Most areas with a water deficit are located near the equator

....

....

⑩ Water use   Grade 3

4. Study **Figure 1**, a map showing water use in the USA in 2010.

(a) Identify the total water use for Texas in 2010. **[1 mark]**

....

(b) Identify a state that had a 9001–22 500 million litre total daily water use in 2010.   **[1 mark]**

....

(c) Using **Figure 1**, suggest **one** reason for the distribution of water use in the USA.   **[2 marks]**

....

....

....

....

AL Alabama			**PA** Pennsylvania
AZ Arizona			**RI** Rhode Island
AR Arkansas	**KS** Kansas	**NE** Nebraska	**SC** South Carolina
CA California	**KY** Kentucky	**NV** Nevada	**SD** South Dakota
CO Colorado	**LA** Louisiana	**NH** New Hampshire	**TN** Tennessee
CT Connecticut	**ME** Maine	**NJ** New Jersey	**TX** Texas
DE Delaware	**MD** Maryland	**NM** New Mexico	**UT** Utah
FL Florida	**MA** Massachusetts	**NY** New York	**VT** Vermont
GA Georgia	**MI** Michigan	**NC** North Carolina	**VA** Virginia
ID Idaho	**MN** Minnesota	**ND** North Dakota	**WA** Washington
IL Illinois	**MS** Mississippi	**OH** Ohio	**WV** West Virginia
IN Indiana	**MO** Missouri	**OK** Oklahoma	**WI** Wisconsin
IA Iowa	**MT** Montana	**OR** Oregon	**WY** Wyoming

Key: 0–9000 | 9001–22 500 | 22 501–45 000 | 45 001–90 000 | 90 001–171 000

Figure 1 Water withdrawals in million litres per day

 **Made a start**   **Feeling confident**   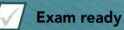 **Exam ready**

Impacts of water insecurity

② Quick quiz

1. Is the statement below true or false?
Circle the correct answer.

Water insecurity can result in people drinking water contaminated by agricultural or industrial waste.

true   false

2. Complete the sentence.

Water .... is the control and development of resources to use less water and prevent ....

3. Complete the sentence.

Water is essential for ....

processes. Water .... can lead

to a reduction in manufacturing.

4. Is the statement below true or false?
Circle the correct answer.

Possible environmental impacts of water insecurity include violence and conflict.

true   false

② Seasonal food consumption and urban farming   Grades 1–2

1. Identify the statement that best describes water insecurity.
Shade **one** circle only.   **[1 mark]**

> The word 'insecurity' suggests that water supply is not reliable.

A Where a place has a surplus supply of fresh water to meet the needs of its people   ○

B Where a place has a limited supply of fresh water but is able to meet the needs of its people   ○

C Where a place has a limited supply of fresh water and is unable to meet the needs of its people   ○

Exam focus 📌

Be careful to shade in the correct number of circles in multiple choice questions as you will not receive any marks if you shade in too many.

⑤ Impacts of water insecurity   Grade 4

2. Outline how water insecurity can lead to waterborne diseases.   **[2 marks]**

....

....

....

....

3. Explain how water insecurity can impact upon food production.   **[2 marks]**

....

....

....

....

Increasing water supply

② Quick quiz

1. Complete the sentence by ticking **one** of the options below.

   Desalination is...

   ☐ ...an inexpensive way of supplying fresh water.

   ☐ ...the removal of salt and minerals from a water source to produce fresh water.

2. Complete the sentence.

   A dam is built to increase ....

3. Complete the sentence.

   Water transfer involves the movement of water from an area of surplus to an area of ....

4. Give **one** disadvantage of using dams and reservoirs.

   ....

⑤ Water transfer schemes   Grade 4

1. Explain **two** advantages of a large-scale water transfer scheme you have studied.   **[4 marks]**

   One advantage of the South-North Water Diversion Project in

   China is that it has the potential to increase agricultural and

   industrial supplies by providing ....

   ....

   ....

Named example

You should have studied a large-scale water transfer system for your exam. You can use this as an example in longer questions, even if you are not specifically asked to. For example, for question **2** you could use the South-North Water Diversion Project in China.

⑩ Increasing water supply   Grades 5–9

2. Explain how water supplies can be increased sustainably.   **[6 marks]**

   ....
   ....
   ....
   ....
   ....

Exam focus

For this 6-mark **explain** question you need to offer several different strategies and give clear reasons to show how each one increases water supplies. To get the top marks, you should describe several strategies for increasing water supply in detail.

   ....
   ....
   ....
   ....
   ....
   ....
   ....
   ....

☐ **Made a start**   ☐ **Feeling confident**   ☐ **Exam ready**

Sustainable water options

② Quick quiz

1. Which of the following are methods of water conservation? Tick **two** boxes.

☐ Installing a low-flush toilet   ☐ Washing a few garments at a time

☐ Installing a smart meter   ☐ Brushing teeth with the tap running

2. Which sustainable water option is the industrial reuse of treated water? Tick **one** box.

☐ Recycling   ☐ 'Grey' water   ☐ Water conservation   ☐ Groundwater management

3. Complete the sentence using a word from the box.

Water stress can be tackled by using

.... water management options.

electric	reserved	sustainable

⑤ Managing water supply   Grade 4

1. Outline how the use of 'grey' water can be used to manage water sustainably. **[2 marks]**

Grey water reduces the total water use of households by ....

....

2. Which **one** of the following statements best describes groundwater management? Shade **one** circle only. **[1 mark]**

A It involves monitoring the quality of water under the surface of the Earth. ◯

B It involves the reuse of treated water. ◯

C It involves adopting ways to limit waste in the home. ◯

⑤ Water conservation and sustainability   Grade 4

3. Explain **one** advantage of a local scheme in a lower income country (LIC) or newly emerging economy (NEE) to increase sustainable supplies of water. **[2 marks]**

....

....

....

....

....

> **Named example** 📍
>
> You need to know an example of a local scheme to increase sustainable supplies of water in an LIC or NEE. You should know the name of the scheme, its location, and several advantages and disadvantages.

4. Explain how recycling can help the sustainable management of water. **[2 marks]**

....

....

....

....

Distribution of energy

② Quick quiz

1. Complete the sentence.

   If a country's energy demand exceeds

   .... then there is an energy

   ....

2. Complete the sentence.

   Global energy consumption has been rising at around

   .... per cent per year since 2000.

4. Complete the sentence.

   China and the USA have the ....

   level of energy consumption in the world.

3. Give **one** factor which affects energy supply.

   ....

⑩ Energy supply and consumption   Grade 3

1. Suggest reasons for variations in energy availability.   **[2 marks]**

   ⚐ *Some countries have invested in technology. This has enabled them*

   to extract ....

   ....

> Revise all the factors affecting energy supply. You could be asked about physical, political or financial factors as well as technological factors.

2. Suggest why global energy consumption is increasing.   **[3 marks]**

   ....

   ....

   ....

   ....

⑤ Energy insecurity   Grade 2

3. Study **Figure 1**, a map showing the percentage of people in African countries who had no access to electricity in 2014.

   (a) Identify the percentage of the population of Ghana that did not have access to electricity in 2014.
   [1 mark]

   ....

   (b) Identify a country where more than 75% of the population did not have access to electricity.
   [1 mark]

   ....

   (c) Using **Figure 1**, describe the pattern of access to electricity in Africa in 2014.   **[2 marks]**

   ....

   ....

   ....

Morocco, Tunisia, Western Sahara, Algeria, Libya, Egypt, Central African Republic, Senegal, The Gambia, Mauritania, Mali, Niger, Sudan, Eritrea, Guinea-Bissau, Chad, Djibouti, Sierra Leone, Guinea, Ghana, Nigeria, Cameroon, South Sudan, Ethiopia, Somalia, Burkina Faso, Togo, Benin, Gabon, DR Congo, Uganda, Kenya, Rwanda, Burundi, Liberia, Côte d'Ivoire, Equatorial Guinea, Tanzania, Malawi, Republic of the Congo, Angola, Zambia, Mozambique, Madagascar, Zimbabwe, Botswana, Namibia, South Africa, Swaziland, Lesotho

% population without access to electricity
- >75%
- 50–75%
- 25–49%
- <25%

Figure 1

 Made a start   **Feeling confident**   **Exam ready**

Impacts of energy insecurity

② Quick quiz

1. Is the statement below true or false? Circle your answer.

Changes to farming methods have caused an increased reliance on energy to produce food.

true   false

2. Name **one** possible environmental impact of drilling for oil in the Arctic.

....

....

⑤ Energy insecurity   Grade 3

1. Which **one** of the following statements best describes energy security?
Shade **one** circle only.   **[1 mark]**

A Energy security is where a place has lots of valuable fossil fuels.   ◯

B Energy security is where a country sells lots of electricity to other countries to expand their economy.   ◯

C Energy security is uninterrupted availability of energy sources at an affordable price.   ◯

⑤ Impacts of energy insecurity   Grade 4

2. Suggest how energy insecurity impacts upon food production.   **[2 marks]**

Uncertainty over the availability of fossil fuels in the future may reduce

the availability of food products

....

....

> **Exam focus**
>
> For Paper 3: Geographical applications, you need to understand the relationships between different geographical issues, such as how energy resources can affect food production.

3. Explain how energy insecurity can lead to conflict.   **[2 marks]**

....

....

....

....

....

4. Explain how energy insecurity can impact upon industrial output.   **[2 marks]**

....

....

....

....

....

Increasing energy supply

⏱ ② Quick quiz ⌛✓

1. Is the statement below true or false?
   Circle the correct answer.

   Renewable resources are finite.

   true   false

2. Which **one** of the resources below is the odd one out?
   Circle your answer.

   oil   solar energy   coal

3. Is nuclear power a renewable or non-renewable resource?

   ....

4. Name **two** renewable energy sources.

   ....

⏱ ⑤ Managing energy supply   Grade 4 ✓

1. Explain **one** advantage and **one** disadvantage of the extraction of fossil fuels in a named location.   **[4 marks]**

   🚩 **Advantage**   <u>Oil extraction from the Athabasca tar sands in Alberta,</u>
   <u>Canada, provides employment for hundreds of thousands of people.</u>
   ....

   Disadvantage ....
   ....
   ....
   ....

   📍 **Named example**

   You need to know an example to show how the extraction of a fossil fuel has advantages and disadvantages. In your exam, you can give a detail from your example, such as the number of people employed, and then use this to illustrate a more general advantage of fossil fuels extraction, such as benefits to the economy.

⏱ ⑩ Sustainable energy supplies   Grades 5–9 ✓

2. Explain how sustainable energy supplies can be increased.   **[6 marks]**

   ....
   ....
   ....
   ....
   ....
   ....
   ....
   ....
   ....
   ....
   ....

   📌 **Exam focus**

   Write clearly and neatly so that your answers can be read easily. If you have time in your exam, check back through your answers to make sure they make sense.

✓ **Made a start**   ✓ **Feeling confident**   ✓ **Exam ready**

Sustainable energy options

② Quick quiz

1. Name **one** way individuals can reduce their energy consumption at home.

....

2. Give **one** way that an individual can reduce their carbon footprint.

....

3. Complete the sentences below.

Improved technology is reducing fuel .... for air, rail and road transport. Engine efficiency has

improved, meaning vehicles can travel .... per litre of fuel than they used to.

⑤ Sustainable use of energy   Grade 4

1. Outline how the reduction of individual carbon footprints can be used to manage energy sustainably.   **[2 marks]**

> In this question, you need to explain **how** these options help the sustainable use of energy, not just list ways to reduce carbon footprints.

🚩 People can reduce their carbon footprint by using public transport. .....

....

....

....

2. Suggest how energy can be used sustainably in the workplace.   **[2 marks]**

....

....

....

⑩ Energy supply options   Grades 5–9

3. Using an example you have studied, to what extent can local renewable energy schemes provide sustainable energy?   **[6 marks]**

....

....

....

....

....

....

....

....

....

....

Named example 📍

You need to know an example of a local renewable energy scheme to provide sustainable supplies of energy in a lower income country (LIC) or newly emerging economy (NEE). For example, the Belo Monte Dam in Brazil was constructed because there is a huge demand for energy there, and the many rivers in the country provide opportunities for generating hydroelectricity. Make sure you revise the example you studied in your course.

Making geographical decisions

⏱ **15** **Responses to natural hazards** | Grades 5–9 ☑

1. To what extent is a country's recovery from a natural hazard based on its level of development? **[9 marks]**

🚩 Recovery from a natural hazard is almost always slower in lower income countries (LICs) and newly

emerging economies (NEEs) than in higher income countries (HICs). This is due to a combination of factors,

....

....

....

> A balanced argument and evaluation of your argument is needed to support your judgement.

....

....

....

....

....

....

....

> Continue your answer on your own paper.

....

Exam focus 📌

For a 9-mark question, it is important to finish your answer with a conclusion, in which you summarise your main points and make a balanced judgement that answers the question.

Exam focus 📌

Refer to named examples and case studies you have learned about to support your points. Try to use specific facts, figures and statistics from your examples.

> Think about how the short-term and long-term responses available to LICs and NEEs are different to those available to HICs.

⏱ **10** **Reducing the development gap** | Grades 5–9 ☑

2. Evaluate the effectiveness of debt relief and fairtrade in reducing the development gap. **[6 marks]**

....

....

....

....

....

....

....

....

> Continue your answer on your own paper.

☑ **Made a start** ☑ **Feeling confident** ☑ **Exam ready**

Fieldwork

On Paper 3, you will need to answer questions about the individual fieldwork enquiry that you have carried out. You will also need to answer questions about an 'unfamiliar' fieldwork scenario. You will be expected to apply your knowledge and use your geographical skills in fieldwork questions.

(10) Data collection, presentation and risk assessments

Grades 5–7 ✓

1. Study **Figure 1**, a photo of a section of river. Suggest why this environment would make a good location for a geographical enquiry. **[2 marks]**

....

....

2. Suggest **one** enquiry question that could be investigated in the area around **Figure 1**. **[2 marks]**

....

....

3. Suggest **one** piece of advice that should be given to students to reduce potential risks when carrying out a physical geography enquiry. **[1 mark]**

....

Figure 1

(15) Extended answer question

Grades 5–9 ✓

4. With reference to your data collection methods, results and conclusions, to what extent could one of your geographical enquiries be improved? **[9 marks]**
[+ 3 SPaG marks]

Title of fieldwork enquiry: ....

....

....

....

....

....

....

....

....

....

....

....

Exam focus 📌

You need to include specific details from all stages of your enquiry in your answer. Revise key points and conclusions from your fieldwork before your exam, so you can mention them in your answers.

Exam focus 📌

This question has 3 marks for **spelling**, **punctuation** and **grammar**, so you should use specialist vocabulary carefully. When you have finished writing, read through and check your answer for any mistakes.

This question asks you to consider all aspects of your fieldwork enquiry critically. Write about what you would change about any part of your fieldwork enquiry if you did it again, and give reasons.

Continue your answer on your own paper.

✓ Made a start ✓ Feeling confident ✓ Exam ready

Practice paper: Living with the physical environment

Time: 1 hour 30 minutes

For this paper you must have: a ruler, a pen, a pencil, an eraser.

The total number of marks for this paper is 88.

Answer **all** questions in Section A and Section B.

Answer **two** questions in Section C.

Section A The challenge of natural hazards

Answer **all** questions in this section.

Question 1 The challenge of natural hazards

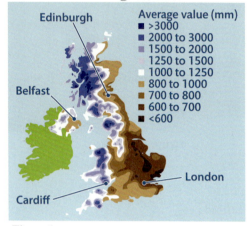

Figure 1

Figure 2

Study **Figure 1**, a map showing the distribution of annual rainfall in the UK from 1981–2010.

01.1 State the average range of annual rainfall for Cardiff. **[1 mark]**

01.2 Describe the distribution of annual rainfall shown in **Figure 1**. **[2 marks]**

01.3 Suggest **one** reason for the distribution of annual rainfall in **Figure 1**. **[2 marks]**

Study **Figure 2**, a photograph of the aftermath of a tropical storm.

01.4 Using **Figure 2**, suggest **two** possible impacts of the tropical storm. **[4 marks]**

Study **Figure 3**, a cross-section diagram of a destructive plate margin.

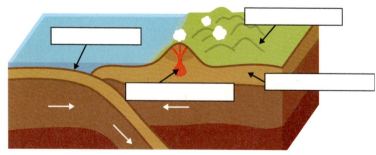

Figure 3

01.5 Complete **Figure 3** by writing the correct label in each box. Choose from the labels below. **[2 marks]**

| oceanic plate | continental plate | fold mountains | magma chamber |

01.6 Which **one** of the following is a climatic hazard? Shade **one** circle only. **[1 mark]**

   A Earthquake ◯   **B** Flooding ◯   **C** Tsunami ◯   **D** Volcanic eruption ◯

01.7 Suggest why people choose to live in areas at risk from a tectonic hazard. **[6 marks]**

01.8 Explain how volcanoes are formed at a constructive plate margin. **[4 marks]**

01.9 Assess the extent to which monitoring and prediction can reduce the risks from a tectonic hazard. **[9 marks]**

[+3 SPaG marks]

Section B The living world
Answer **all** questions in this section.

Question 2 The living world
Study **Figure 4**, a world map showing the distribution of tropical rainforests.

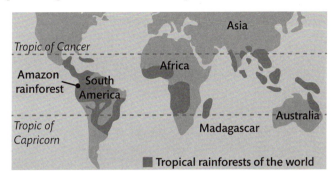

Figure 4

02.1 In which continent is the Amazon rainforest found? Shade **one** circle only. [1 mark]

A North America ⊚ **B** South America ⊚ **C** Australia ⊚ **D** Brazil ⊚

02.2 Give **one** characteristic of tropical rainforests. [1 mark]

02.3 Describe the distribution of tropical rainforests shown in **Figure 4**. [2 marks]

Study **Figure 5**, a photograph of a commercial palm oil plantation on an area of cleared tropical rainforest.

Figure 5

02.4 Using **Figure 5**, describe and explain how commercial plantations cause deforestation. [6 marks]

Study **Figure 6**, an example of a tropical rainforest climate graph, and **Figure 7**, a table showing the rainfall in a rainforest in January.

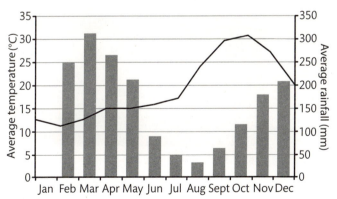

Figure 6

Month	Rainfall (mm)
January	230 mm

Figure 7

02.5 Using the data table in **Figure 7**, complete the bar for rainfall in January. [1 mark]

02.6 What is the total annual rainfall shown in **Figure 6**? Shade **one** circle only. [1 mark]

A 2600 mm ⊚ **B** 2000 mm ⊚ **C** 1750 mm ⊚ **D** 2400 mm ⊚

02.7 State the month that experiences the highest level of rainfall. [1 mark]

02.8 Outline **two** opportunities created from living in hot deserts or cold environments. **[2 marks]**

02.9 For a hot desert environment or cold environment you have studied, to what extent are there challenges and opportunities for people living in that environment? **[9 marks]**

Section C Physical landscapes in the UK
Answer **two** questions from the following:
Question 3 (Coasts), Question 4 (Rivers), Question 5 (Glacial).
Shade the circles below to indicate which **two** optional questions you will answer.

Question 3 ◯   Question 4 ◯   Question 5 ◯

Question 3 Coastal landscapes in the UK
Study **Figure 8**, a diagram showing features of coastal erosion.

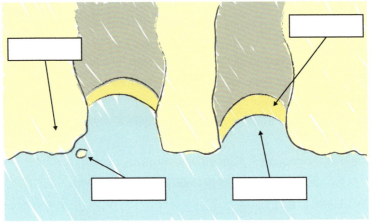

Figure 8

03.1 Complete **Figure 8** by writing the correct label in each box. Choose from the labels below. **[2 marks]**

bay	headland	beach	stack

03.2 Complete the following sentences. **[3 marks]**

The direction of longshore .... is determined by the .... The swash of the wave brings the sediment onto the beach at an acute angle, whilst the .... drags the sediment back into the sea. This process continues moving sediment along the coastline.

Study **Figure 9**, a photograph of a coastal landform.

Figure 9

Figure 10

03.3 Using **Figure 9**, explain how processes cause the formation of the landform. **[4 marks]**

Study **Figure 10**, a photograph showing an example of a coastal management scheme in Burnham-on-Sea.

03.4 With the help of **Figure 10**, explain how hard engineering is used to protect coastlines from the effects of coastal erosion. **[6 marks]**

Question 4 River landscapes of the UK

04.1 Complete the following sentences.   **[3 marks]**

There are two types of erosion that occur along the course of a river channel: .... erosion and

lateral erosion. One process of erosion is the force of the water hitting the bed and .... of the

river, leading to increased pressure, which wears away the river channel. This is called .... action.

Study **Figure 11**, a diagram showing the cross-section of a meander.

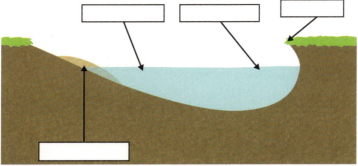

Figure 11

04.2 Complete **Figure 11** by writing the correct label in each box. Choose from the labels below.   **[2 marks]**

| slip-off slope | fastest flow | slowest flow | river cliff |

Study **Figure 12**, a photograph of a river landform on the River Dove, in Derbyshire.

Figure 12

Figure 13

04.3 Using **Figure 12**, explain the processes that cause the formation of the landform.   **[4 marks]**

Study **Figure 13**, a photograph showing river management in the Elan Valley in Wales.

04.4 With the help of **Figure 13**, explain how hard engineering is used to protect settlements from
the effects of river flooding.   **[6 marks]**

Question 5 Glacial landscapes in the UK

05.1 Complete the following sentences.   **[3 marks]**

Deposition occurs during melting or glacial .... Rotational ....

occurs when a glacier becomes lubricated from summer meltwater, which causes it to slide downhill. The

glacier transports .... within it.

Study **Figure 14**, a diagram showing a glacial moraine.

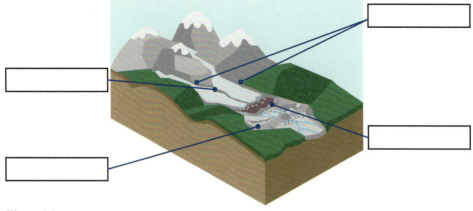

Figure 14

05.2 Complete **Figure 14** by writing the correct label in each box. Choose from the labels below.　　**[2 marks]**

terminal moraine	lateral moraine	medial moraine	recessional moraine

Study **Figure 15**, a photograph of a glaciated landform in Snowdonia National Park.

Figure 15

Figure 16

05.3 Using **Figure 15**, explain how processes cause the formation of the tarn lake.　　**[4 marks]**

Study **Figure 16**, a photograph showing a glaciated upland area in Snowdonia National Park.

05.4 With the help of **Figure 16**, explain how tourism can impact on glaciated landscapes.　　**[6 marks]**

Practice paper: Challenges in the human environment

Time: 1 hour 30 minutes

You must have: a ruler, a pen, a pencil, an eraser.

The total number of marks for this paper is 88.

Answer **all** questions in Section A and Section B.

Answer Question 3 and **one** other question in Section C.

Section A Urban issues and challenges

Answer **all** questions in this section.

Question 1 Urban issues and challenges

Study **Figure 1**, showing the global percentage of people living in urban areas since 1960.

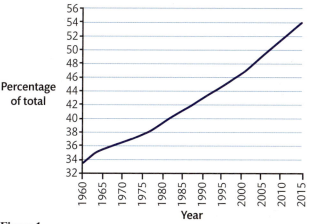

Figure 1

01.1 What percentage of people lived in urban areas in 2000? **[1 mark]**

01.2 Calculate the difference between the percentage of people living in urban areas in 1980 and in 2015. **[2 marks]**

01.3 Outline **one** factor that may have caused the rapid growth of urban areas. **[2 marks]**

Study **Figure 2**, an Ordnance Survey map extract showing part of the city of Manchester.

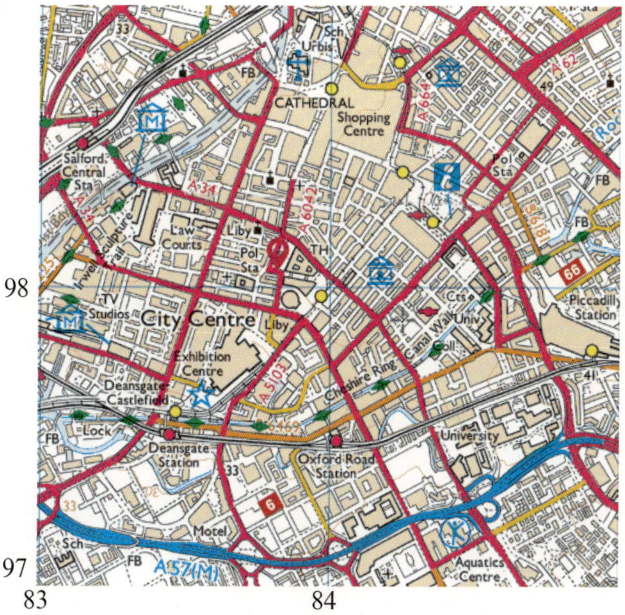

Figure 2 An extract of an OS map of Manchester. Scale 1:25000

01.4 Give the four-figure grid reference for Deansgate station. **[1 mark]**

01.5 What is the six-figure grid reference for the cathedral? Shade **one** circle only. **[1 mark]**

    A 839979 ◯   **B** 838984 ◯   **C** 848979 ◯   **D** 838987 ◯

01.6 Greater Manchester is one of the UK's most populous areas. Suggest **one** human factor that may affect population density. **[1 mark]**

01.7 Using **Figure 2** and an example of a major city you have studied, discuss the challenges created by urban growth. **[6 marks]**

Study **Figure 3**, showing information for a sustainable management strategy in a London suburb.

Sustainable management strategies in BedZED, Beddington

BedZED is a sustainable urban development, created on a brownfield site in Beddington, London. The development provides 100 affordable eco-houses with the vision to be a carbon neutral community. The strategies installed and promoted by the developers are as follows:

1. Solar panels installed on the roofs of the houses, which contribute directly to the electricity supply
2. South-facing windows to capture heat
3. Wind cowls to encourage air circulation
4. Car-share club to encourage car sharing amongst the community
5. Built near public transport

Figure 3

01.8 Suggest why the location of BedZED is an advantage to its management strategies. **[2 marks]**

01.9 Using **Figure 3** and your own knowledge, discuss the strategies used to manage resources and transport sustainably. **[6 marks]**

01.10 Evaluate the effectiveness of an urban regeneration project(s) you have studied. **[9 marks]**
[+ 3 SPaG marks]

Section B The changing economic world

Answer **all** questions in this section.

Question 2 The changing economic world

Figure 4

Study **Figure 4**, a world map showing the global distribution of Gross National Income (GNI) per capita.

Gross National Income (GNI) is the total value of a country's goods, services and overseas investments, divided by the number of people in that country.

02.1 Complete **Figure 4** using the information below. **[1 mark]**

The GNI per capita for Finland is $43,400.

02.2 What is the GNI per capita for Ghana? **[1 mark]**

02.3 Suggest the limitation of using only GNI per capita to judge a country's level of development. **[2 marks]**

02.4 Explain **two** causes of uneven development. **[4 marks]**

Study **Figure 5**, a graph showing the UK's changing employment structure from the 1930s onwards.

02.5 Calculate the difference between the tertiary and secondary sectors in 2011. **[1 mark]**

02.6 What was the percentage of primary sector employment in 1981? **[1 mark]**

02.7 Suggest reasons for the UK's changing employment structure shown in **Figure 5**. **[4 marks]**

02.8 Suggest **two** strategies used to solve the regional differences within the UK. **[4 marks]**

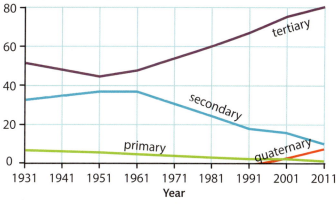

Figure 5

Figure 6

Study **Figure 6**, a photo showing construction work in Mumbai, India.

02.9 Using **Figure 6**, outline **one** impact of rapid economic growth. **[2 marks]**

02.10 Using **Figure 6** and your own knowledge, evaluate the effectiveness of strategies used to reduce the development gap. **[9 marks]**

Section C The challenge of resource management

Answer Question 3 and **either** Question 4 **or** Question 5 **or** Question 6.

Question 3 The challenge of resource management

Study **Figure 7**, a graph showing the total primary energy consumption in the UK by type of fuel from 1970 to 2014.

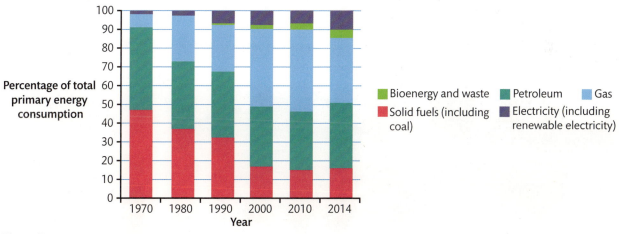

Figure 7

03.1 In 2010, what percentage of the total primary energy consumption was supplied using petroleum? Shade **one** circle only. **[1 mark]**

   A 31% ◯   **B** 41% ◯   **C** 51% ◯   **D** 61% ◯

03.2 Using **Figure 7**, describe the changes to the UK's primary energy consumption since 1970. **[3 marks]**

03.3 Give **two** reasons for the increase in electricity generated by renewables. **[2 marks]**

03.4 Outline **one** environmental issue associated with the exploitation of energy resources. **[2 marks]**

03.5 To what extent should there be less reliance on fossil fuels and more reliance on renewable resources to provide the UK's energy? **[6 marks]**

Answer **either** Question 4 (Food) **or** Question 5 (Water) **or** Question 6 (Energy).
Shade the circle below to indicate which optional question you will answer.
Question 4 ◯   Question 5 ◯   Question 6 ◯

Question 4 Food

Study **Figure 8**, a global map showing the prevalence of undernourishment in the population.

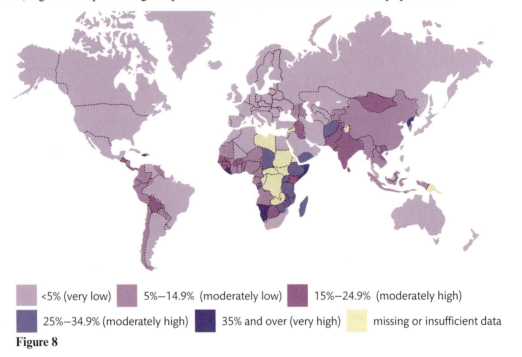

▣ <5% (very low)	▣ 5%–14.9% (moderately low)	▣ 15%–24.9% (moderately high)
▣ 25%–34.9% (moderately high)	▣ 35% and over (very high)	▣ missing or insufficient data

Figure 8

04.1 What is the percentage of undernourishment for India? Shade **one** circle only.   **[1 mark]**
    A 5%–14.9% ◻   **B** 15%–24.9% ◻   **C** <5% ◻   **D** 25%–34.9% ◻

04.2 Describe the distribution of the countries that have a high percentage of undernourishment.   **[2 marks]**

04.3 Suggest **one** way in which rising food prices can affect food security.   **[2 marks]**

04.4 Using an example you have studied, discuss the advantages and disadvantages of strategies used
to increase food security.   **[6 marks]**

Question 5 Water

Study **Figure 9**, a global map showing global water stress in 2013.

▣ low (<10%)
▣ low to medium (10–20%)
▣ medium to high (20–40%)
▣ high (40–80%)
▣ extremely high (>80%)

Figure 9

05.1 What is the percentage of water stress for India? Shade **one** circle only.   **[1 mark]**
    A 80% ◻   **B** 10–20% ◻   **C** 40–80% ◻   **D** 20–40% ◻

05.2 Describe the distribution of countries that have a high percentage of water stress.   **[2 marks]**

05.3 Suggest **one** way conflict can cause water insecurity.   **[2 marks]**

05.4 Using an example you have studied, discuss the advantages and disadvantages of strategies
used to increase water security.   **[6 marks]**

Question 6 Energy

Study **Figure 10**, a global map showing the percentage of electricity generation from renewable sources.

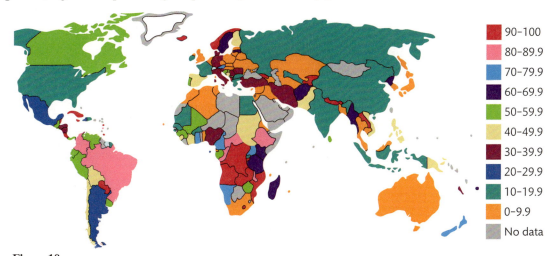

Figure 10

06.1 What is the percentage of electricity generation from renewable sources for India? Shade **one** circle only.

[1 mark]

    A 10–19.9% ◯   **B** 90–100% ◯   **C** 20–29.9% ◯   **D** 50–59.9% ◯

06.2 Describe the distribution of countries that have a low percentage of electricity generation from renewable sources. [2 marks]

06.3 Suggest **one** way conflict can cause energy insecurity. [2 marks]

06.4 Using an example you have studied, discuss the advantages and disadvantages of strategies used to increase energy security. [6 marks]

Resources for Practice paper: Geographical applications

To be issued to students 12 weeks before the exam

This booklet contains three resources as follows:

1. Figure 1 – Evidence for climate change
2. Figure 2 – The effects of climate change
3. Figure 3 – Responding to climate change

Exam focus 📌

For Paper 3: Geographical applications, you will be issued with a pre-release resources booklet 12 weeks before your exam. You can study the booklet and make notes on it, but you will be given a clean copy to take into the exam.

Figure 1 Evidence for climate change

The amount of carbon dioxide in the air has increased significantly in recent years.

The annual global temperature is increasing and provides evidence for climate change, commonly known as global warming. The year 2010 was the warmest year globally since records began in 1880, with some scientists predicting that temperatures will continue to rise rapidly. In 2010, the annual global combined land and ocean surface temperature was 0.94 °C above the 20th century average. This was the 39th consecutive year (since 1977) that the yearly global temperature was above average.

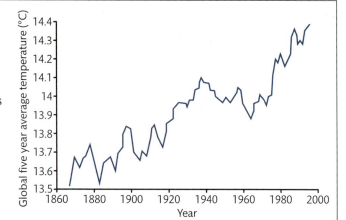

The percentage of carbon dioxide in the atmosphere is rising at a very high rate, due to human activity. Burning fossil fuels, such as coal and oil, releases carbon dioxide into the atmosphere. Scientists believe that the rise in the concentration of CO_2 enhances the greenhouse effect. Deforestation also contributes to the significant rise in the concentration of CO_2 in the atmosphere. This is because the trees felled no longer absorb CO_2 and the carbon stored in the trees is released into the atmosphere as CO_2 when they are burned.

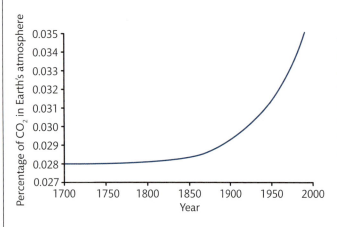

Figure 2 The effects of climate change

The rising temperature in the Arctic is causing the ice to melt, with ice coverage reducing by 4% every 100 years. This is affecting polar bears' ability to find food; the retreating ice means they have less time to hunt for prey and they go without food for longer than usual.

Over time, this leads to female bears struggling to find food for their cubs, resulting in fewer surviving into adulthood.

Figure 2 The effects of climate change continued

Rising sea temperatures are increasing the bleaching of coral reefs, with scientists predicting that by 2050, 98% of reefs around the world will be affected. One consequence is that the changes to the aesthetic appeal of the coral damages tourism, resulting in a loss of revenue. This can threaten the livelihoods of communities. Corals provide shelter and food for the wider ecosystem and the damage being caused to these reefs is having a wider impact on fish species.

Rising sea levels could threaten the lives of people living in low-lying coastal areas, such as Bangladesh and the Maldives. Changes in sea levels have disrupted fishing activity around the islands of the Maldives, which is one of the most important industries in supplying food for the local people. The Asian tsunami in 2004 caused devastating damage to the islands, affecting the tourism industry. In the future, rising sea levels could result in the loss of the islands.

Figure 3 Responding to climate change

Carbon Capture

Avoiding dangerous climate change is still possible, but will cost more than twice as much if we don't utilise carbon capture and storage (CCS). **Intergovernmental Panel on Climate Change, 2014.**

In 2014, SaskPower launched its $1.5 billion carbon capture and storage project at its Boundary Dam Power Station in Canada. The project aims to reduce carbon emissions by 90%, turning the burning of coal, a fossil fuel, into a cleaner energy generation process. Once the gas has been captured it is injected into a nearby oil field to remove oil from the ground, using a technique called enhanced oil recovery. Carbon is then stored deep underground in a saline aquifer. The Boundary Dam is one of 22 carbon capture projects worldwide that are either operational or under construction to reduce global emissions.

Key benefits of the project at Boundary Dam include:

1. The project has prevented over two million tonnes of carbon dioxide from reaching the atmosphere.

2. The amount of carbon captured by the power station is equivalent to taking 250,000 cars off the road every year.

3. The company plans to reduce its emissions by 40% from 2005 levels by 2030.

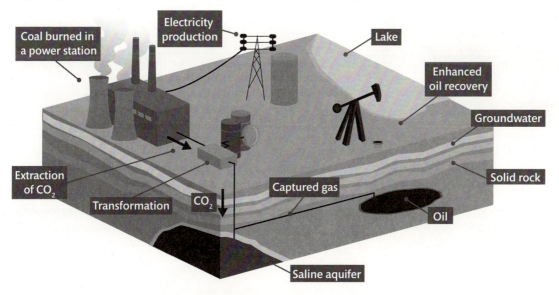

Problems associated with carbon capture and storage are:

4. There are concerns about potential leakage of carbon during the capture process.

5. The storage of carbon has increased the potential of leaks into groundwater supplies, which would increase the acidity of the water and make it undrinkable.

Practice paper: Geographical applications

Time: 1 hour 15 minutes

For this paper you must have: a clean copy of the resources booklet.

You must have: a ruler, a pen, a pencil, an eraser.

The total number of marks for this paper is 76.

Answer **all** questions.

Exam focus 📌

Make sure you have studied the Resources on pages 100–101 before starting this paper. You will need to refer back to the Resources for questions **1**, **2** and **3**.

Section A Issue evaluation

Answer **all** questions in this section.

Study **Figure 1**, 'Evidence for climate change', in the resource booklet.

01.1 Describe the changing pattern in the average temperature of the planet since 1860. **[2 marks]**

01.2 Outline **one** reason for the increasing amount of carbon dioxide in the air. **[2 marks]**

01.3 With the help of **Figure 1**, suggest why reducing carbon emissions is important to the long-term sustainability of our planet. **[6 marks]**

01.4 Suggest **one** environmental effect of increasing average temperature. **[2 marks]**

Study **Figure 2**, 'The effects of climate change', in the resource booklet.

02.1 What is the current estimated percentage of coral reefs affected by coral bleaching? Shade **one** circle only. **[1 mark]**

    A 96% ◯   **B** 98% ◯   **C** 90% ◯   **D** 92% ◯

02.2 'Climate change is having significant impacts on people and the environment.'

Use **Figure 2** and your own knowledge to discuss this statement. **[6 marks]**

Figure 4 The rise of the electric car

Electric vehicles have become increasingly popular, with more than 155,000 plug-in vehicles on the road in 2018, compared with 3500 in 2013. This increase has been possible due to improvements in vehicle designs and a greater number of charging stations.

Advantages
• Electrically powered engines mean they don't emit toxic gases or smoke into the environment.
• Electric cars reduce noise pollution in cities. They can travel long distances without the high acceleration.

Disadvantages
• The reduced noise pollution can create a hazard, due to some people being unable to hear the vehicles approaching.
• The current buying market for electric cars is limited, resulting in higher prices than petrol or diesel vehicles.

Study **Figure 3**, 'Responding to climate change', in the resource booklet, and **Figure 4**, 'The rise of the electric car', above.

03.1 Suggest **one** way the type of transport shown in **Figure 4** might affect climate change. **[2 marks]**

03.2 Explain how carbon capture and storage pose potential risks to the environment and human health. **[4 marks]**

03.3 Do you think carbon capture and storage and electric vehicles are effective strategies in the response to climate change? Shade **one** circle.

    Yes ◯   No ◯

Use evidence from the resource booklet and your own knowledge to explain your answer. **[9 marks]**

[+3 SPaG marks]

Section B Fieldwork

Answer **all** questions.

Figure 5

Study **Figure 5**, a photograph of part of an urban area.

04.1 Identify **two** data collection techniques that could be used to carry out a geographical fieldwork investigation in the area shown.

[2 marks]

Technique 1: ....

Technique 2: ....

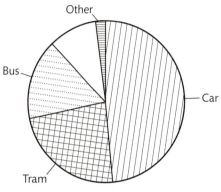

Figure 6 Types of transport used between 9 a.m. and 9.30 a.m.

Study **Figure 6**, a pie chart showing the results of data collected by students who were carrying out a geographical enquiry about congestion problems in a town centre. The sample size was 100.

04.2 Complete the pie chart for 4 pedestrians and 6 cyclists.　[2 marks]

04.3 Suggest **two** alternative methods of presenting the information shown in **Figure 6**.　[2 marks]

As part of an enquiry collecting primary physical geography data, a student measured pebble size every 5 metres along a beach transect. The results are shown in **Figure 7**.

Study **Figure 8**, a scattergraph of the pebble sizes against the distance from the sea.

Sample	Distance from the sea in metres	Pebble size in centimetres
1	5	1
2	10	3
3	15	3
4	20	6
5	25	9
6	30	10
7	35	13
8	40	17
9	45	8
10	50	22

Figure 7

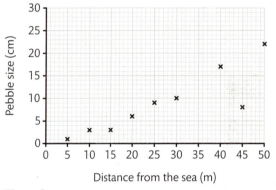

Figure 8

04.4 Complete **Figure 8** using the data for sample 7 in **Figure 7**.　[1 mark]

04.5 Explain **one** disadvantage of the data collection technique in **Figure 7**.　[1 mark]

04.6 Using the data in **Figure 7**, calculate the mean of the pebble size data. Show your working.　[2 marks]

04.7 Describe the relationship between distance from the sea and pebble size shown in **Figure 8**.　[4 marks]

04.8 Suggest **two** pieces of advice that should be given to students in order to reduce potential risks when carrying out a **physical** geography enquiry.　[2 marks]

05.1 State the title of your fieldwork in which **physical** geography data were collected.

　　Title of fieldwork enquiry: ....

　　Explain the choice of the location(s) used for your geographical enquiry.　[2 marks]

05.2 Justify **one** primary data collection method used in your **physical** geography enquiry.　[3 marks]

05.3 State the title of your fieldwork in which human geography data were collected.

　　Title of fieldwork enquiry: ....

　　Assess how effective your chosen techniques were in representing the data collected in this enquiry.　[6 marks]

05.4 With reference to your methods, results and conclusions, assess how **one** of your geographical enquiries could be improved.　[9 marks]

[+3 SPaG marks]

Answers

Page 1 Defining natural hazards

Quick quiz: 1. drought; tropical storm
Questions:
1. *Any of the following*: an earthquake; intense flooding; landslide following a hurricane or tornado. **2.** population; growing **3.** C
4. If a natural hazard occurs where there is a large population, more people and homes will be at risk of damage. For instance, the number and size of cities in seismically active areas is growing, meaning that the risk of damage from earthquakes and volcanoes will be higher and the damage will affect more people.
5. Urbanisation means that an increasing proportion of the global population live in urban areas, which tend to be densely populated and are often in areas at risk from natural hazards. This increases the natural hazard risk.

Page 2 Distribution of earthquakes and volcanoes

Quick quiz: 1. oceanic
2. inner core, outer core, mantle, crust
3. thick; older; thin; younger
Questions:
1. A; C
2. Volcanoes in South America are distributed in a linear pattern along the western coastline.
3. Convection currents in the Earth's mantle cause tectonic plates to move. This generates huge amounts of energy at plate margins, which causes earthquakes and volcanic eruptions.
4. Intraplate earthquakes are earthquakes that occur away from plate margins.

Page 3 Plate margin activity

Quick quiz: 1. volcanoes **2.** constructive (island arcs are found at destructive plate boundaries)
Questions:
1. Feature X   oceanic crust
      Feature Y   continental crust
2. destructive plate margin **3.** *Any one of the following*: volcanoes; deep sea trenches; fold mountains.
4. Figure 1 shows a destructive plate margin where two tectonic plates are moving towards each other. The denser oceanic plate (X) is subducted beneath the less dense continental plate (Y). When the build-up of intense friction and pressure between the two plates is released, it causes shock waves and can result in earthquakes. The plates stick together then jolt suddenly and release. This gives off seismic waves causing an earthquake in the area above on the Earth's surface.

Page 4 Named example: Earthquake effects

Quick quiz: 1. ...something that happens as a direct result of a tectonic hazard. **2.** true
Questions:
1. *Any one of the following*: people being killed; buildings being destroyed; damage to roads and other infrastructure.
2. Earthquakes can trigger other hazard events, such as landslides and tsunamis, leading to further damage and risk to life. *Other answers might include: people being left homeless; rural villages being destroyed by landslides; avalanches; tremors felt in other countries; the economic impact of rebuilding.*
3. *Example answer*: One reason is that more developed countries are able to invest in building earthquake-proof buildings, which reduces the risk of collapse. This limits the

cost of the damage and saves lives. For instance, Ecuador has invested in more earthquake-resistant buildings than Nepal and, consequently, suffered less damage in the 2016 earthquake than Nepal did in the 2015 earthquake.
4. *Example answer*: Primary effects of earthquakes include loss of life, injuries and destruction of buildings. For example, in the 2015 Nepal earthquake the primary effects included more than 8000 deaths, many thousands more injuries, and more than 600 000 damaged or destroyed buildings. These significantly disrupt the lives of survivors. For example, many people will grieve the loss of loved ones and many will be homeless.

Page 5 Named example: Earthquake responses

Quick quiz: 1. Underground mining; Movement of tectonic plates
2. ...an immediate response.
Questions:
1. *For example*: **Response 1:** One long-term response to an earthquake is the NGO Oxfam working with local communities to raise awareness of good hygiene practices to limit the spread of waterborne diseases. **Response 2:** Countries work with organisations such as the United Nations to establish a national building code to improve the structural design of buildings. *Other answers might include: rebuilding water supply systems; preparing an action plan to manage the risk of future tectonic hazards.*
2. *Example answer*: Following the 2016 earthquake in Ecuador, the government worked with NGOs such as Oxfam, to educate local communities about good hygiene practices in an effort to reduce the spread of waterborne diseases such as cholera. Alongside educational support, Oxfam co-ordinated the rebuilding of water supply systems in the municipality of San Vicente to provide clean water for local people. These measures limited the risk of further loss of life from the secondary effects of the earthquake. After the 2015 earthquake in Nepal, however, the long-term responses were focused upon improving the country's readiness for future tectonic hazards, including working with the UN to establish a national building code to improve the structure of buildings.
3. *Example answer*: **Response 1:** One immediate response is initiating a state of emergency, as the government of Nepal did in 2015, when they deployed the army to help with the search and rescue effort. **Response 2:** A second immediate response is the provision of emergency medical services, such as those provided by the International Medical Corps after the earthquake in Ecuador in 2016. *Other answers might include: the UN and NGOs organising international aid; neighbouring countries assisting with rescue operations; implementation of mobile hospitals; police dog units used in search and rescue.*

Page 6 Living with natural hazards

Quick quiz: 1. true **2.** increased
Questions:
1. Areas with volcanic activity can provide opportunities to generate geothermal energy. For example, approximately 66% of Iceland's energy comes from geothermal sources.
2. Hazardous areas have diverse landscapes that attract tourists, which provides employment opportunities for local people working as tour guides.
3. *Example answer*: I agree with the statement because, although the unpredictability of natural hazards creates risks for those people who live near them, they also provide significant benefits.

One benefit of living in hazardous areas is the fertile soils created by volcanic ash in the area surrounding a volcano. Minerals in volcanic ash, such as nitrogen and potassium, create rich soils leading to increased crop yields. This in turn increases farmers' income. However, the risk of farming near a volcano is that crops can be completely destroyed by an eruption, which can also endanger the homes and lives of farmers.

The awe and wonder of volcanic landscapes attracts many visitors each year from around the world, creating employment opportunities in tourism and boosting the local economy. Local people can get jobs as tour guides, enabling them to earn an income for their families. However, although this is undoubtedly a benefit of living in a hazardous area, the employment is often seasonal and therefore income can be inconsistent, especially during and immediately after an eruption, when tourist numbers decrease.

Another reason people may choose to live in hazardous zones is that in more developed countries many people view the risk from hazards such as earthquakes as low, because improved technology can provide advanced warning and many buildings are now earthquake-resistant. For example, in San Francisco, many buildings are equipped with shock absorbers and counterweights that help to keep the building stabilised during tremors, reducing the possibility of it collapsing. Monitoring by scientists and early warning systems also mean people can be evacuated away from some hazards, reducing the risk.

Overall, hazardous zones provide many benefits for people, whose quality of life may decrease if they moved away from an area of natural hazard risk, and the infrequency of natural hazard events and improvements to building design and evacuation plans mean that the risks are outweighed by the benefits.

Page 7 Reducing the risk

Quick quiz: 1. *Any two of the following:* thermal imaging equipment; GPS, tiltmeters and laser beams; seismometers; chemical sensors. **2.** earthquakes
Questions:
1. Hazard mapping can be used to predict the location of future hazards, which allows town and city planners to manage the construction of infrastructure so that high risk areas are not built upon. For example, the purple area in **Figure 1** would not be built upon as the tectonic hazard risk is very high in this area.
2. Earth embankments can be built around a volcano, which can divert the flow of magma away from settlements. *Other answers might include: exclusion zones; earthquake-proof buildings; earthquake drills.*
3. Earthquake-proof buildings can reduce the risk by using features such as shock absorbers and counterweights, which reduce the impact of tremors on buildings and help prevent their collapse.
4. Seismometers record activity in and around volcanoes. Increased activity can provide scientists with advanced information about the likelihood of an earthquake or volcanic eruption, allowing warnings to be issued and evacuation procedures to be implemented.

Page 8 General atmospheric circulation model

Quick quiz: 1. Hadley cell
2. Cool air that moves back towards the equator
Questions:
1.

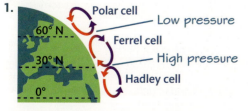

2. warmer; polar
3. At the equator, warm air rises, causing low pressure. The rising air moves north and south, and cools to form Hadley cells. At latitudes of 30° north and 30° south, the cooled air sinks, creating areas of high pressure. The air loses moisture and sinks, meaning that clouds cannot form and so hot deserts occur; temperatures get very high, and at night temperatures drop quickly.

Page 9 Distribution of tropical storms

Quick quiz: 1. …June and November.
2. …South Pacific or the Indian Ocean. **3.** weaker
4. water vapour

Questions:
1. Hurricanes can form in the Pacific and Atlantic Oceans, moving west. Hurricanes in the Atlantic Ocean move towards Central America and the south-eastern coast of North America.
2. Typhoons begin in the north-west Pacific Ocean, moving north-westerly towards the south-east of Asia, hitting countries such as Thailand.
3. *Any two of the following:* the Coriolis effect creating spin; warm air rising from the ocean; smaller storms joining to form a tropical storm; lots of water vapour in the atmosphere; cooling water vapour, which forms cumulonimbus clouds; warm sea temperatures.
4. Tropical storms occur between the tropics because they require certain conditions to form. Within 5° of the equator the Coriolis effect is not normally strong enough, and further away from the tropics the sea temperature is normally too cold (below 26.5 °C).

Page 10 Causes and features of tropical storms

Quick quiz: 1. lower **2.** cumulonimbus
Questions:
1.

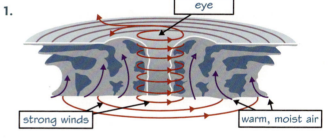

2. A; C
3. Rising sea temperatures from climate change can affect the distribution of tropical storms, with storms occurring outside their current distribution. Climate change can also increase the intensity of tropical storms, making them more destructive.
4. low; rises

Page 11 Named example: Tropical storms

Quick quiz: 1. (a) surges **(b)** Primary effects
Questions:
1. *Any one of the following:* deaths and injuries; trees uprooted; damage to infrastructure.
2. Effect 1: Destruction to homes leaving people homeless
Effect 2: Pollution of water sources leading to widespread illnesses
3. *Example answer:* One long-term response to tropical storms is the establishment of livelihood support programmes to help people earn an income whilst they are dealing with the damage caused by the storm. For example, after Typhoon Haiyan, livelihood support programmes gave cash grants for debris clearance and recycling.
4. *Example answer:* One immediate response to Typhoon Haiyan was the distribution of emergency kits from donor countries, such as the UK and Canada, to help with the provision of necessities such as food and shelter.

5. NGOs can help to educate people in planning to reduce the risk from tropical storms in the future. NGOs can also respond to a tropical storm by sending volunteers to help survivors rebuild their homes.

Page 12 Planning for tropical storms

Quick quiz: 1. Figure 1: protection; **Figure 2:** planning
Questions:
1. Strategy 1: Advising people how to prepare for and respond to a tropical storm. **Strategy 2:** Governments can use planning policies that prevent construction of buildings in high-risk areas. *Other answers might include: installing early warning systems, which help to reduce the death toll; advising people in areas at risk of tropical storms to stock up on food and water.*
2. Countries can construct storm shelters, which include features that make them strong enough to survive tropical storms. For example, they are made of reinforced concrete, metal shutters cover the windows and they are built off the ground. *Other answers might include: high sea walls can be built to protect against storm surges.*
3. A major benefit of monitoring a storm is that large numbers of people who may be affected can be informed of the storm's progress through TV and the internet, and receive live updates as it develops. This informs people how likely they are to be affected, and warns them to prepare for the storm, or to evacuate if necessary. Another benefit of using storm tracking systems is that storms can be detected long before they reach land, which gives the maximum possible time for people to prepare.
4. LICs cannot afford to invest in expensive storm-tracking technology, and may not have the technology to effectively communicate information to people who may be affected by a tropical storm, especially in remote rural areas.

Page 13 Weather in the UK

Quick quiz: 1. ...millimetres (mm) or metres (m)
2. *Any three from the following:* very cold weather; flash floods; heavy snowfall; storms; heavy rainfall; heatwaves leading to periods of drought.
Questions:
1. The map shows that there is a greater number of properties in the centre that are at risk of flooding. For example, Southwark is one of the boroughs at highest risk of flooding with more than 50 000 properties at risk.
2. C **3.** south-east; drought

Page 14 UK extreme weather events

Quick quiz: 1. true
Questions:
1. An economic impact of the flood shown in **Figure 1** might be the cost of damage to cars, homes and businesses. A social impact of the flood is the disruption to road travel, which will affect businesses, schools and access to services. *Other economic impacts might include: businesses being forced to close temporarily and thus losing income; reduction in productivity if people are unable to get to work. Other social impacts might include: children may not be able to go to school; people may be trapped in their houses; emergency services may find it difficult to reach people; risk of injury to people trying to get through the floodwater; risk to health posed by waterborne diseases; trauma of home being damaged and having to move out whilst it is repaired.*
2. An example of an extreme weather event is flooding. This can be caused by prolonged periods of heavy precipitation.
3. Dredge rivers regularly, to prevent the build-up of sediment in the river channel. This increases the river's carrying capacity and reduces flood risk.
4. *Example answer:* The Somerset floods of 2013–2014 had a negative economic impact on the agricultural industry, with more than 11 500 hectares of farmland underwater for over 15 days. This led to loss of income for farmers because the land took time to recover. It also caused a wider scale social impact, as food supply was affected. Another economic impact was that many local businesses were forced to close, as premises were damaged by floodwater. This contributed towards an average loss of income of £17 352 per business. The floods also led to the disruption of people's daily lives, with many roads and bridges closed due to the floodwaters, increasing commuter time by 40%, preventing access to shops for food and stopping pupils from attending school. This resulted in many individuals and families relying on each other for water and food provision during this difficult time.

Page 15 Climate change evidence

Quick quiz: 1. true **2.** ice cores or pollen analysis
Questions:
1. One piece of evidence is retreating glaciers: scientists suggest they are retreating at double the rate of a decade ago in areas like the Himalayas. *Other answers might include: rising global temperatures; rising ocean temperatures; the ice sheets in Greenland and Antarctica are shrinking; many areas of the northern hemisphere have had record low levels of snow cover.*
2. wider; climate
3. 56 mm (*allow range of +/- 2 mm*)
4. 17 mm (*allow range of +/-2 mm*)
5. Although sea levels decreased slightly at various points, for instance after 1998 and in 2010–2011, the trend shown in **Figure 1** is that sea levels increased steadily, at a rate of approximately 3 mm per year.

Page 16 Causes of climate change

Quick quiz: 1. (a) true **(b)** false
2. *Any two of the following:* coal; gas; oil.
Questions:
1. The farming of livestock, especially cows, causes the release of methane, which is one of the main greenhouse gases that contribute to the greenhouse effect.
2. In the past, extreme periods of volcanic activity caused large volumes of carbon dioxide (CO_2), a greenhouse gas, to be released into the atmosphere. This changed the atmosphere and caused global warming.
3. *Any two of the following:* every 100 000 years, the Earth's orbit changes from elliptical, which causes warmer periods, to less elliptical, which causes cooler periods; the Earth's tilt on its axis varies over a period of 41 000 years, and a larger tilt causes more extreme seasons; the Earth sometimes wobbles whilst spinning, which can affect the severity of the seasons in one hemisphere compared with the other.
4. When humans burn fossil fuels, such as coal and oil, it causes an increased release of carbon dioxide into the atmosphere. This rise in the concentration of carbon dioxide enhances the greenhouse effect, causing temperatures to rise. Another human activity that contributes to climate change is deforestation, as carbon stored in the trees is released into the atmosphere as carbon dioxide when the trees are felled or burned. *Other answers might include: deforestation reduces the number of trees that absorb carbon dioxide; agriculture can result in large amounts of greenhouse gases being release into the atmosphere.*

Page 17 Managing climate change

Quick quiz: 1. ...mitigation. **2.** ...mitigation.
Questions:
1. *Example answer:* Low-lying islands, such as the Maldives, have houses built on stilts to protect the residents from rising sea levels. This means that their houses will be protected from the rising water. *Other answers might include: agricultural systems; water supply management.*

2. *Example answer:* International agreements, such as the 2015 Paris Agreement, are important because countries that sign up must commit to reducing CO_2 production and rises in global temperatures, mitigating the effects of climate change.

3. Afforestation will contribute to reducing the concentration of CO_2 in the atmosphere because trees absorb carbon dioxide during photosynthesis.

4. As temperatures increase, crops will need more water. Farmers will need to adapt their irrigation methods to make them more efficient, for example by using drip feed irrigation systems. This will allow them to save water whilst maintaining crop yields.

Page 18 Exam skills: Using examples and case studies

1. *Example answer:* Both immediate and long-term responses are significant when reacting to an earthquake or volcanic eruption and can, to a large extent, determine how severe the impact is on the environment and on people in the area affected by the hazard.

Long-term responses to an earthquake can be particularly significant in countries that are vulnerable to regular earthquake hazards. After the 2015 earthquake in Nepal, long-term responses included the non-governmental organisation GeoHazards International setting up the Kathmandu Valley Earthquake Risk Management Project, and the United Nations working with Nepal to create a building code to increase the earthquake resistance of buildings constructed in the future. Although these long-term responses will have a positive effect in the event of future earthquakes, Nepal is a lower income country, and as such was unable to invest the amount of money in long-term responses that a higher income country might have. As such, the long-term responses were comparatively basic.

Immediate responses are essential to minimise the damage after an earthquake. For example, after the 2015 earthquake in Nepal, over £87 million of international aid was raised through the Nepal Earthquake Flash Appeal. This helped provide food, medicine and shelter for those effected. The Nepalese army, with assistance from neighbouring countries, carried out search and rescue operations and saved many lives. This had a much more significant impact on the Nepalese people than the long-term responses. However, the immediate responses relied heavily on the assistance of the international community. These took longer to be implemented, resulting in more deaths than in a higher income country where instant action could be taken. In conclusion, both long-term and immediate responses are important when reacting to an earthquake. However, the immediate response has a much more significant impact on the lives and well-being of the people affected, particularly in lower income countries where there is less money to be invested in long-term responses.

2. *Example answer:* Many scientists believe that the natural causes of climate change, like orbital changes, solar cycles and volcanic activity, do not account for the recent increasing global temperatures. The concentration of CO_2 in the atmosphere has been increasing at a rapid rate since the Industrial Revolution, which has changed the composition of the Earth's atmosphere. This has caused the enhanced greenhouse effect, in which less solar radiation is reflected back into space. Rather, it is trapped within the atmosphere and warms the planet.

The rising concentration of CO_2 in the atmosphere is caused entirely by human activity, fuelled by an increasing global population, all of whom require food, water and energy. The rising demand for energy and food has resulted in an increased rate of deforestation, which contributes to greenhouse gas emissions because the trees are no longer absorbing carbon dioxide during photosynthesis, and trees release carbon dioxide when they are felled or burned. Use of fossil fuels for transport is also a major cause of greenhouse gas emissions.

In summary, evidence suggests that the growing world population – more than seven billion people – has increased the demand for key resources, which is increasing human activities such as agriculture and deforestation, which release greenhouse gases and exacerbate climate change.

Page 19 A small-scale ecosystem

Quick quiz: 1. false   **2.** ...all the non-living things.
3. an animal that eats other animals
4. *Any two of the following:* tropical rainforest; desert; polar; coniferous forest; deciduous forest; taiga forest; tundra; savannah; grassland; Mediterranean.
Questions:
1. producers   **2.** water beetle / pond snail
3. A food web shows how all the food chains in a particular ecosystem interact.
4. A decline in the population of carnivores, such as frogs, could lead to a rapid increase in the herbivores, their prey, which in turn will put pressure on the producers. An increase in carnivore numbers would reduce numbers of herbivores, and thus the producers would increase in number.
5. C

Page 20 Tropical rainforests

Quick quiz: 1. the shrub level   **2. (a)** hot; wet
(b) drip-tip; buttress
Questions:
1. B
2. The climate in tropical rainforests is always hot and wet, which means the growing season lasts all year round. The abundance of vegetation supports a wide range of insect, bird and mammal species.
3. The climate of tropical rainforests is hot and humid all year round, with average daily temperatures of 28 °C. Another characteristic is that rainfall levels are high, with downpours occurring daily. These typically amount to over 2000 mm per year. *Other answers might include: local tribes use food and resources sustainably and live in harmony with the environment; tropical rainforests are made up of four key layers: the forest floor, under canopy, the canopy and the emergent layer; that tropical rainforests have very high levels of biodiversity; because of the heavy rainfall, nutrients are leached from the soil, so decomposing organic matter on the forest floor is the main source of nutrients; plants are adapted to the climate, for example drip-tip leaves and buttress roots.*

Page 21 Case study: Deforestation

Quick quiz: 1. *Any one of the following:* increased soil erosion and risk of flooding; loss of habitat; increased carbon dioxide emissions; fewer trees to act as carbon 'sinks'.
2. *Any three of the following:* population growth; settlement / urbanisation; energy development; HEP; mineral extraction; road building; logging; commercial farming; subsistence farming.
Questions:
1. (a) 2012   **(b)** 1.3 million hectares
(c) Figure 1 shows an overall increase in the rate of deforestation and the hectares of tree cover lost from Indonesia's primary forests. Tree cover loss increased from 0.7 million hectares in 2001 to 1.5 million hectares in 2014.
2. *Example answer:* Deforestation is caused by commercial agriculture, particularly growing crops such as palm oil. For instance, in Indonesia palm oil plantations have caused large areas of rainforest to be cleared using a method called 'slash and burn'. This removes the trees and leaves fertile ash, which helps

the newly planted crop to grow. Subsistence agriculture can also lead to deforestation, because many farmers cut down forest or burn it in order to clear land for growing food for their families. Beef production can also lead to deforestation, as rainforest may be cleared to provide grazing land for cattle.

Page 22 Sustainable rainforests

Quick quiz: 1. (a) biodiversity   **(b)** National parks
(c) reduction; protection
Questions:
1. Soft trekking guides tourists on a specifically allocated trail, which prevents damage to wildlife and plant species.
2. It helps local people to make a living from the rainforest, which gives them a powerful incentive to protect it.
3. The use of selective logging is vital to the conservation of tropical rainforests, as it ensures that countries can benefit economically from their rainforest without causing unsustainable deforestation. Because trees are only cut down when they are over a certain height, and only one or two species in an area are logged, the amount of trees harvested at one time is reduced and most of the forest is left alone.
4. To promote the global trade of tropical timber from sustainably managed forests.

Page 23 Characteristics of hot deserts

Quick quiz: 1. 250 mm   **2.** 20%   **3.** The Sahara
4. *Any one of the following:* habitat degradation; less predictable rainfall; rising temperatures.
Questions:
1. A
2. One of the characteristics of hot deserts is that they are very dry, with average precipitation levels of less than 250 mm per year. As well as being dry, hot deserts are hot during the day, with temperatures exceeding 50°C all year round. *Other answers might include: desert soils are dry, thin and sandy. Deserts also have very low levels of biodiversity. Deserts can get very cold at night, with temperatures below 0°C.*
3. Camels have adapted to surviving without water for several months. They sweat as little as possible and store fat in their humps, so that it does not insulate their whole bodies and overheat them. *Other answers might include: lizards and snakes only active during the early morning and shelter under sand / rocks; owls and bats more active during the night when temperatures are lower.*
4. Cacti are xerophytes, which means they are able to store water for long periods of time. They do not have leaves, which reduces water loss through transpiration. *Other answers might include: desert lilies have heat and drought resistant bulbs and seeds; phreatophyte plants have long roots to reach the water table.*

Page 24 Case study: Opportunities and challenges in a hot desert

Quick quiz: 1. minerals   **2.** tourism   **3.** farming
4. renewable / solar
Questions:
1. The harsh landscape of deserts such as the Sahara in North Africa make it difficult for people to develop settlements because of difficulties building key infrastructure. Another challenge of living in places like the Sahara is water scarcity and drought due to the low levels of precipitation during the year. This also means that food security is a problem. *Other answers might include: the health risk posed by extreme temperatures in the desert; deserts are also often remote and inaccessible, and may be hundreds of kilometres from the nearest airport and other transport infrastructure.*
2. *Example answer:* One of the opportunities created by hot deserts is a range of tourist activities, for example, the Sahara

Desert is a popular tourist destination for camel trekking and 4×4 tours around Erg Chebbi. This means that the income generated from the activities contributes towards the local economy and provides employment for the local people as tour guides. However, the extremely high temperatures of over 40°C may make it difficult to attract large numbers of tourists. The hot, sunny climate does however provide an ideal opportunity for countries to invest in the generation of renewable solar energy. To support this, Morocco is building the Noor 1 solar power plant which will be the largest in the world, capable of producing an estimated 580 megawatts of energy. This will potentially enable the country to be less reliant on importing energy to meet its demands.
Hot deserts also create farming opportunities, although the hot and dry climate limits the crops that be grown; the main commercial crops in the Sahara are dates and other fruit. Some deserts also have opportunities for mineral extraction. Phosphate, copper and limestone are found in the Sahara desert.

Page 25 Desertification

Quick quiz: 1. hot deserts   **2.** uses less water, so reduces the amount of water wasted
3. (a) overgrazing   **(b)** removal of fuel wood / deforestation
Questions:
1. The intensive growing of crops to meet the demands of a growing population means that the soil is unable to recover each year. This leads to it losing its nutrients and becoming infertile, making it vulnerable to erosion.
2. Climate change is creating hotter and drier conditions in more places. This increases the likelihood of desertification.
3. Appropriate technology involves the use of agricultural techniques that are suited to the knowledge, resources and income of local people, such as putting rocks around fields to trap water, increasing crop yields. This reduces the likelihood of over-cultivation, one of the main causes of desertification. *Other answers might include: water and soil management; tree planting.*

Page 26 Characteristics of cold environments

Quick quiz: 1. Northern Hemisphere
2. *Any two of the following:* polar bears, bears, wolves, arctic foxes, seals, whales, arctic hares, reindeer, moose.
3. Antarctic
Questions:
1. In **Figure 1**, you can see that mosses and lichens grow close to the ground in cold climates such as the Arctic Tundra. This protects them from the wind. *Other answers might include: lichens and some plants grow very quickly in spring when temperatures rise; lichens do not need soil to grow and can survive cold temperatures; Arctic poppies have a hairy stem to help retain heat.*
2. Biodiversity in cold environments tends to be low because the extreme cold and low levels of precipitation make it difficult for many organisms to survive and breed.
3. spring; habitat
4. In cold environments, average winter temperatures range from −40°C in the tundra biome to −60°C at the South Pole. The tundra has a short summer season of 50 days. The tundra biome contains two soil layers: the active layer and the permafrost layer. *Other answers might include: low levels of biodiversity; soils in cold environments are mostly thin and infertile.*

Page 27 Case study: Opportunities and challenges in tundra

Quick quiz: 1. permafrost
2. *Any two of the following:* mineral extraction; tourism; fishing; energy.

Questions:

1. (a) *Example answer:* In Siberia, the landscape and wildlife provide opportunities for tourism, such as outdoor pursuits and wildlife tours. *Other answers might include: energy generation; mineral extraction.*
(b) Food production can be limited by extremely cold climates, so it can be a challenge to provide sufficient food resources to the inhabitants.
2. C
3. *Example answer:* Extreme temperatures in cold environments like the Arctic Tundra mean people require thermal clothing to be able to cope. Traditional tribes, like the Nenets, use reindeer skin as an insulator.

Page 28 Conservation of cold environments

Quick quiz: 1. melting; rise
2. *Any of the following:* maintaining the fragile environment; protecting wildlife; promoting sustainable use of resources.
3. true
Questions:
1. Cold environments provide a home for many indigenous tribes who rely on animals for their food, clothing and tools. If these environments were not protected wilderness areas, changes could result in their traditional way of life being affected, jeopardising their survival. *Other answers might include: the intrinsic value of cold environments; their high air and water quality; the opportunities for recreation; they are fragile, and the animals and plants that live there would not survive changes to the environment.*
2. International agreements provide clear guidelines to establish strategies and agreements to best protect the environment. For instance, the Antarctic Treaty is a commitment to monitoring and regulating scientific research and tourism.
3. In many cold environments, economic development using technology is vital for the future existence of nearby towns and cities, and for the extraction of resources for global consumption. This leads to construction that can pose threats to these fragile environments. An example is the Trans-Alaskan Pipeline. To overcome the potential risk of permafrost melting, the pipeline was built on stilts to hold it off the ground, allowing wildlife species to continue to migrate across the landscape.

Page 29 Landscapes in the UK

Quick quiz: 1. Ben Nevis   **2.** River Severn
3. north-west   **4.** upland
Questions:
1. Upland landscapes are suitable for sheep farming because the soil tends to be thin and infertile, so unsuitable for crop farming. Also, there are many areas of steep relief, which many breeds of sheep are well-adapted to.
2. *Any one of the following:* fertile soils; wide, flat plains; much closer to sea level.
3. (a) B   **(b)** A   **(c)** 167721

Page 30 Waves and weathering

Quick quiz: 1. mechanical
2. ...how far the wave has travelled.
Questions:
1.

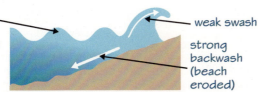

tall waves with short wavelength

weak swash

strong backwash (beach eroded)

2. Constructive waves have a strong swash and a weak backwash. *Other answers might include: they have a long wavelength; they have a low wave height; they cause the deposition of material.*
3. *Any two of the following:* the size of the fetch; how long the wind has been blowing; how fast the wind is blowing.
4. When acidic rainwater falls on rocks, over time this leads to a chemical reaction that breaks down some types of rock, including limestone.
5. When rainwater falls into the cracks of rocks and the temperature falls below zero, the water freezes. The frozen water expands, exerting pressure on the surrounding rock. Over time, the repeated freezing and thawing causes the rock to break apart. *Other answers might include: heating and cooling; biological weathering.*

Page 31 Coastal erosion landforms

Quick quiz: 1. ...waves compressing pockets of air in cracks in a cliff.
2. 1 fault / crack;   **2** cave;   **3** arch;   **4** stack
3. Abrasion is the action of rock particles carried by waves being hurled at a cliff, causing pieces to break off. Attrition is the action of rocks colliding with each other, causing them to become smoother and rounded.
Questions:
1. bays; resistant
2. Destructive waves attack the base of a cliff through the sheer force of the waves compressing air into small cracks within the rock, a process known as hydraulic action. Over time, the continued action of the erosion process causes a wave-cut notch to develop. Continued erosion of the wave-cut notch creates an overhang, which eventually is unable to support its own weight and so collapses under the force of gravity, forming a wave-cut platform. The collapsed material is further eroded by the process of attrition, which smoothes the wave cut platform over time.
3. Destructive waves hit the headland, eroding lines of weakness, or faults, in the rock through the force of the waves compressing air into small cracks within the rock, a process known as hydraulic action. Over time, continued erosion of a fault causes it to grow, eventually forming a small cave. Further erosion of the cave through hydraulic action and abrasion causes the cave to increase in size until it breaks through the headland to form an arch. A combination of hydraulic action and abrasion undercutting the bottom of the arch, and mechanical and chemical weathering affecting the top of the arch, increases the size of the archway. Eventually the arch is unable to support its own weight and collapses due to gravity, forming a column of rock known as a coastal stack. Continued erosion and weathering causes the stack to decrease in size, forming a stump.

Page 32 Mass movement and transportation

Quick quiz: 1. Solution   **2.** rolling
3. ...the rotational collapse of permeable rocks.
4. oblique angle
Questions:
1. (a) the prevailing wind   **(b)** the swash
2. Rock falls happen due to the movement of rock fragments under the influence of gravity, which is increased by the process of mechanical weathering.
3. D

Page 33 Coastal deposition landforms

Quick quiz: 1. *For example:* marram grass ; beach grass; bent grass
2. longshore   **3.** lagoon   **4.** low-energy constructive wave

Questions:

1. Sand dunes are formed from onshore winds blowing sediment to the back of the beach. The sediment is initially deposited around an obstruction, leading to the formation of embryo dunes. Foredunes are formed over time from the growth of vegetation, such as marram grass, which helps to stabilise the dunes. The decomposition of plants provides nutrients, allowing a wider range of plant species to colonise the dunes, causing the dunes to get larger.

2. Figure 1 shows a coastal bar, which is a deposition of sand extending across a bay with a lagoon behind it. A coastal bar forms from the deposition of sediment over time. Firstly, the prevailing wind causes the swash of the waves to carry sediment up the beach at an angle. The backwash then drags the sediment back down the beach at a right angle under the force of gravity. This moves the sediment in a zig-zag pattern along the coastline, in the process of longshore drift. Where there is a sudden change in direction of the coastline, longshore drift continues to transport the material in the same direction, and the transported material is deposited offshore because of the waves losing energy. Over time, sediment continues to be deposited and builds up to form a spit. When the spit reaches another headland, it forms a bar, the landform shown in **Figure 1**, which stretches all the way across a bay.

Page 34 Hard engineering

Quick quiz: 1. a sea wall

2. (a) *One of the following:* it is comparatively cheap; it is quick to implement.

(b) *One of the following:* they are unattractive; they have a short lifespan as they can rust and break.

Questions:

1. Groynes can help to reduce the effects of longshore drift but they can deprive beaches of material further along the coast. *Other answers might include: they reduce space on the beach for recreational activities; they are unattractive.*

2. A

3. *Example answer:* Hard engineering strategies are used to manage coastlines because they are effective at preventing erosion. They are large, solid artificial structures which are designed to absorb or reflect wave energy. For example, sea walls are made of concrete, which is dense and hard-wearing, and they are often curved, which helps to reflect wave energy away from the coastline. At Walton-on-the-Naze, a sea wall was built to protect the buildings nearest the sea. Climate change may cause sea levels to rise and stormy weather to increase, which may mean that some of the hard engineering strategies may not be as effective at managing coastal erosion in the future.

Page 35 Soft engineering

Quick quiz: 1. Beach reprofiling – The transfer of sediment from the lower to the upper beach

Dune regeneration – The artificial creation of new dunes or the restoration of existing dunes

Managed retreat – Allowing the sea to flood in designated areas of the coast

2. *Any one of the following:* it's a sustainable approach; it creates salt marsh habitats; it is comparatively cheap; it doesn't interfere with natural processes.

Questions:

1. B

2. One benefit of using sand dune regeneration is it can contribute towards maintaining a diverse natural environment, which helps to support a wide variety of wildlife. *Other answers might include: It protects buildings on the seafront; in the short term, it is cheaper than hard engineering coastal management strategies.*

3. One of the costs of using managed retreat is that it can result in areas of coastal land being flooded, which can be expensive for the local government due to the possibility of having to compensate local residents for the loss of buildings or farmland. *Other answers might include: it can be unpopular with locals; saltwater can have a negative impact on existing ecosystems.*

Page 36 Coastal management

Quick quiz: 1. *One of the following:* tourism, farming, transport (e.g. railway), energy (nuclear power and oil refineries) **2.** *One of the following:* managed retreat; dune regeneration/stabilisation. **3.** Soft; hard

Questions:

1. (a) 265 236

(b) 2.5 km

2. *Example answer:* The coastal management scheme being implemented at Walton-on-the-Naze in Essex is important because it is an area that relies on tourism to support the local economy. However, in recent years it has experienced very high rates of erosion (2 metres per year). This is because the cliffs consist of London Clay and Red Crag, which are both less resistant rock types and are vulnerable to erosion. This is the reason for the management scheme of constructing a sea wall to protect the buildings nearest the sea, and the installation of Crag Walk in 2011, which uses rock armour to protect the coastline.

The cliffs at Walton-on-the-Naze are also regularly affected by slumping. To address this, drainage channels have been installed in the cliffs to stop the soil becoming saturated and limit further instability. The coastline is also affected by the process of longshore drift moving sediment away from the beach. This resulted in the installation of groynes to limit the effect of longshore drift and help to retain the beach, which is an important tourist attraction. The coastal management scheme incurred costs for construction and maintenance of the sea defences. However, this was necessary to protect the beaches for tourism, which brings money into the local area.

Page 37 A river profile

Quick quiz: 1. ...a stream that joins the main river.

2. the mouth   **3.** the boundary of the drainage basin.

4. ...wide and deep.

Questions:

1. (a) The cross profile in the upper course is V-shaped, with a narrow and shallow channel. **(b)** At point A in **Figure 1**, the river erodes vertically, whereas at point B the river erodes laterally, causing a wider channel. **(c)** Point C is in the lower course, where the increased volume of water means that lateral erosion and deposition are more significant than vertical erosion, causing the channel to flatten and widen.

2. C

Page 38 Fluvial processes

Quick quiz: 1. (a) vertical erosion   **(b)** vertical erosion and lateral erosion   **(c)** lateral erosion

2. Abrasion – Rock particles carried by the river hit the bed and banks, wearing them down.

Attrition – Rock fragments carried by the river collide with one another, causing them to become smaller and more rounded.

Solution – Soluble rocks, such as limestone, dissolve in the river.

Questions:

1. B   **2.** C

3. Deposition is the action of sediment being dropped due to the river losing energy and being unable to transport it any further.

4. middle and lower course

5. Hydraulic action occurs when waves compress the air in cracks in rocks, causing sections of rock to wear or break off, whereas abrasion is the action of rock fragments scraping against the bed and banks.

Page 39 Fluvial erosion landforms

Quick quiz: 1. upper course   **2.** hydraulic action or abrasion
Questions:
1. Label **A** shows hard resistant rock whereas label **B** represents soft, less resistant rock.
2. Gorges form when waterfalls retreat upstream. This happens because soft rock is eroded faster than the hard rock above, creating an overhang. The overhang collapses due to gravity and the process repeats over time. This causes the waterfall to retreat upstream and form a narrow, steep-sided valley known as a gorge.
3. When a river flows through valley slopes, it follows the path of least resistance. When it comes across more resistant rock, the river will change direction. This leads the river to weave between more resistant rocks, and the fluvial erosion creates interlocking spurs.

Page 40 Fluvial erosion and deposition landforms

Quick quiz: 1. middle   **2.** deposition
3. true; false; true
Questions:
1. (a) Feature **X** is a slip-off slope. Feature **Y** is a river cliff.
(b) A river cliff is formed where the line of fastest flow in a river causes erosion on the outside of a bend.
A combination of hydraulic action compressing the air in cracks in the river bank, the sheer force of the water, and abrasion (rock fragments carried in the water hitting the bank and wearing it down) erodes the bank, undercutting it. Over time, this forms a river cliff.
2.
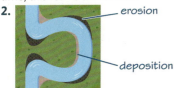
erosion

deposition
3. An estuary is found at a river's mouth. It is a wide body of water with a slow flow where the river flows out into the sea. It is a mixture of fresh water from the river and saltwater from the sea. An example is the Severn Estuary.

Page 41 Flood risk

Quick quiz: 1. (a) social   **(b)** economic
2. (a) impermeable   **(b)** increases   **3.** physical
Questions:
1. The steeper the surrounding slopes of an area, the greater the risk of flooding. This is because water runs off the land much faster, reaching the river channel more quickly.
2. One human factor that increases flood risk is land use. Urban areas have more impermeable surfaces, which increase surface run-off and cause water to reach drains and the river channel faster. Some types of agriculture can leave soil exposed, which also increases surface run-off and therefore increases flood risk.
3. B

Page 42 Flood hydrographs

Quick quiz: 1. river; storm   **2.** slower
Questions:
1. (a) 36–37 cumecs   **(b)** 45 mm   **(c)** 14 hours
2. A flashy hydrograph has a higher peak discharge, a steep rising limb and a shorter lag time, whereas a gentle hydrograph has a lower peak discharge, a less steep rising limb and a longer lag time.
3. If the land surrounding the river consists of lots of steep slopes, the rate of groundwater flow and surface run-off will be faster. This causes the water to reach the channel at a much faster rate than in an area with a gentle gradient. The river discharge increases quickly, creating a flashy hydrograph with a steep rising limb and a shorter lag time. However, if the land surrounding the river has a gentle gradient, surface run-off and groundwater flow will be much slower. This means the water will reach the river more slowly, which will be shown on the flood hydrograph in a less steep rising limb. It will also lead to a longer lag time because the river will take more time to reach peak discharge.

Page 43 Hard engineering

Quick quiz: 1. (a) embankments or levees
(b) straightening
Questions:
1. The use of dams and reservoirs can be effective in providing a source of water, and they can be used to generate hydroelectricity, which can contribute towards reducing reliance on fossil fuels and carbon emissions. However, the costs of using dams and reservoirs to manage rivers includes the fact that they're very expensive to install and maintain. They can also cause farmland downstream to become less fertile.
2. Flood relief channels can be used to direct high water levels to another area. Therefore, they effectively reduce the risk of flooding in a particular area.
3. River straightening to manage rivers allows the water to flow faster, potentially reducing flooding in a particular area. Rivers can also be made wider and deeper when they are being straightened, which increases the amount of water the river channel can hold before it floods.

Page 44 Soft engineering

Quick quiz: 1. (a) flood warnings and preparation
(b) flood plain zoning
Questions:
1. B
2. Flood plain zoning can provide land for alternative uses, such as pasture, farming and parkland. *Other answers might include: it reduces the risk of hospitals, schools and houses being built on land prone to flooding.*
3. One of the benefits of using river restoration to manage a river is that, once the hard engineering strategies have been removed, there are no ongoing maintenance costs. *Other answers might include: reintroducing meanders slows the river flow, which reduces erosion.*
4. One cost associated with using afforestation to manage a river is that it takes a long time for trees to become effective at intercepting precipitation. *Other answers might include: afforestation alone doesn't prevent flooding.*

Page 45 Flood management

Quick quiz: 1. *Any one of the following:* wildlife habitat destroyed; crops and fields underwater for long periods of time; animals drowning; soil and water becoming polluted by sewage.
2. *Any one of the following:* people's lives disrupted; houses uninhabitable; injuries or death; spread of disease; people unable to get to school and work; people trapped and unable to get food.
Questions:
1. One possible short-term impact is damage to properties. A second possible impact is people sustaining injuries.
2. One possible economic effect is that the businesses shown in **Figure 1** will lose money whilst they close for repairs. *Other answers might include: loss of income from tourism; cost of repairs to houses; insurance claims; cost of replacing cars that were washed away.*
3. *Example answer:* One of the environmental issues associated with the management scheme at Boscastle was the disturbance and destruction of some wildlife habitats, impacting on the wider ecosystem of the local area.

4. *Example answer:* Following the 2004 flooding in Boscastle, which caused over £15 million worth of damage to property and resulted in over 150 people being rescued, the local council and the Environment Agency worked to implement a £10 million flood management scheme. One of the long-term responses involved in the scheme was the widening of the river channel near the Riverside Hotel and the car park. This improved the carrying capacity of the River Valency, increasing the time it would take for it to reach peak discharge and thereby reducing flood risk. Second, two new sewer pump stations were installed which were implemented to help divert floodwater, along with a trash screen to prevent debris from blocking the drainage channel, reducing the potential risk of flooding. These measures were necessary to protect people and properties in the village from any future effects of flooding. By spending money to implement the flood management scheme to protect the village, the aim is to prevent any future costs for replacement and rescue.

Page 46 Glacial processes

Quick quiz: 1. pushing loose rock fragments forwards.
2. true   **3.** Cardiff   **4.** striations
Questions:
1. Freeze-thaw weathering occurs when water enters the cracks of rocks. When it freezes, the water expands, which causes pressure on the rock. Over time, the repeated process of freezing and thawing results in the production of angular fragments called scree.
2. Outwash is the sand and gravel deposited by running water when a glacier melts.
3. Rotational slip can occur when a glacier becomes lubricated by summer meltwater, which causes it to slide downhill. The glacier transports material within it.
4. The main deposition happens at the end of the snout of the glacier where the greatest amount of melting takes place.
5. C

Page 47 Glacial erosion landforms

Quick quiz: 1. Arêtes are knife-edged ridges.
2. hollow   **3.** false   **4.** peaks; arêtes
Questions:
1. B   **2.** A   **3.** C   **4.** C
5. *Example answer:* Ribbon lakes, such as Windermere in the Lake District, form when a glacier flows over and erodes softer rock in glacial troughs. The glacier erodes softer rock to a greater depth than hard rock. Following the retreat of the glacier, meltwater collects in the deepened area, forming a ribbon lake.

Page 48 Glacial transportation and deposition landforms

Quick quiz: 1 D   2 A   3 B   4 E   5 C
Questions:
1. (a) terminal moraine   **(b)** The terminal moraine is the pile of deposits carried or pushed by a glacier to form a high ridge at the snout of the glacier.
2. Erratics occur when rocks transported by a glacier are deposited as the glacier loses energy. The glacier leaves behind rocks that look out of place.
3. glacial deposits; direction

Page 49 Economic activities

Quick quiz: 1. lowland areas
2. *Any one of the following:* noise pollution; air pollution; loss of recreation areas; landscape scarring; disturbs wildlife; destroys wildlife habitat.
Questions:
1. The soils in upland glaciated landscapes are rocky and thin, which means they lack the nutrients for crops to grow well. They are suited to pastoral farming, as animals such as sheep are well-adapted to poor grazing and steep slopes.
2. Farmers can diversify by setting land aside for campsites, boosting their income. *Other answers might include: converting farm buildings to holiday rentals.*
3. new business opportunities; employment
4. Tourists can cause traffic congestion, increasing journey time for commuters. *Other answers might include: they may drop litter, disrupt wildlife or contribute to footpath erosion; tourism may lead to pressure to develop an area, which may be unpopular with local people.*
5. It can damage natural habitats in upland areas, as native species of tree may be felled in order to clear land for commercial plantations.

Page 50 Tourism in Snowdonia

Quick quiz: 1. employment.
2. *Any two activities, for example:* hiking, cycling.

Questions:
1. *Example answer:* One strategy that can be used to manage the impact of tourism is to mark specific footpaths through environmentally sensitive areas, such as the maintained paths throughout Snowdonia National Park. This can help to prevent the erosion of local plants. *Other answers might include: use information boards to increase awareness of protecting the environment; reward businesses with funding for being more sustainable.*
2. (a) *Any one of the following:* for the peace and quiet; for the beautiful mountain scenery; to spend time in a wilderness area; to take part in outdoor sports and activities.
(b) It may increase people's interest in, and desire to conserve, wild places and the habitats of the plants and animals that live there.
3. It may provide only seasonal employment for local people, which may not be sufficient for them to be able to live in the area. *Other answers might include: tourism can increase house prices due to the demand for holiday homes, making local people unable to afford to live there.*

Page 51 Exam skills: Levelled response questions

1. Figure 1 shows a waterfall with a plunge pool beneath it. **Figure 1** shows there is a harder, resistant rock overlaying a softer, less resistant rock. Over time, a plunge pool forms as the river flows over rapids. The presence of white, foaming water indicates fast-flowing water where hydraulic action, the sheer force of the water pushing air into the cracks of the rocks, has caused the softer, less resistant rock to erode at a faster rate than the harder, more resistant rock. This process, combined with abrasion, causes an overhang to form. Eventually, the harder rock is unable to support its own weight and collapses due to gravity, forming a waterfall and plunge pool as shown in **Figure 1**. The collapsed material provides further material for erosion and the deepening of the waterfall.
2. Figure 2 shows steep-sided chalk cliffs, which contain wave-cut notches at the base, where the sea meets the cliffs. These types of cliffs are formed through mechanical and chemical weathering processes. Destructive waves attack the base of the cliffs through the erosional process of hydraulic action, which is the force of the waves pushing air and water into the cracks of the rocks, causing them to break apart. Over time, the continued repetition of hydraulic action could cause a wave-cut notch to form. The combination of hydraulic action and abrasion, which is the action of rock fragments being hurled at the cliffs, causes the wave-cut notch to increase in size, forming an overhang. Eventually the overhang will collapse under the force of gravity, forming a steep-sided cliff again. Loose rocks at the base of the cliffs are evidence that this process has occurred.

Page 52 Urbanisation characteristics

Quick quiz: 1. a city with over 10 million inhabitants
2. *Any one of the following:* lack of affordable housing in urban areas; better quality of life in rural areas; more green space in rural areas, retirement.
3. true **4.** rural–urban; urban; rate
Questions:
1. Figure 1 shows that the majority of megacities are located in the northern hemisphere, with more megacities in Asia than any other continent. There are only four megacities in South America, and none in Australasia.
2. One factor that affects the rate of urbanisation is natural increase – when the birth rate is higher than the death rate. Natural increase occurs because urban areas tend to have better healthcare and a larger number of people of reproductive age.
3. rural; urban; reversed
4. HICs already have high levels of urbanisation, whereas in LICs and NEEs, the rate of urbanisation is increasing rapidly as more and more people are moving from rural areas into urban areas to take up jobs in manufacturing. Industries have relocated from HICs to NEEs.

Page 53 Case study: Opportunities in LICs and NEEs

Quick quiz: 1. rural–urban migration
2. *Any one of the following:* access to health services; access to education; access to water supply; access to energy
3. Gross Domestic Product
Questions:
1. *Example answer:* In rural areas of LICs and NEEs, job opportunities are limited, and jobs such as farming, as shown in **Figure 1**, are poorly paid and have poor working conditions. This encourages people to leave rural areas and move to a city such as Mumbai, where there are better paid employment opportunities.
2. *Example answer:* In Mumbai, rapid urban growth has resulted in a rise in the service sector, which has created more employment opportunities for people in a variety of service sector jobs, such as couriers, mechanics, cleaners and hairdressers. *Other answers might include: urban industrial areas can stimulate economic development; small businesses thrive in many urban areas, such as the Dharavi slum in Mumbai, where small businesses generate more than $650 million every year.*
3. *Example answer:* In Mumbai's slums, rapid growth has created close-knit communities, where many residents conduct their chores together. For example, they meet to wash clothes and to socialise at the communal washing areas. *Other answers might include: healthcare is improving; access to education is improving; a greater percentage of the urban population are gaining access to energy and clean water.*

Page 54 Case study: Challenges in LICs and NEEs

Quick quiz: 1. squatter **2.** *Any one of the following:* dangerous working conditions; no taxation; poor pay; no job security.
Questions:
1. *Example answer:* In Mumbai, waste disposal is a challenge because rivers are becoming polluted, leading to waterborne diseases, like cholera. *Other answers might include: slums and squatter settlements; the challenge of providing clean water and energy supplies; the pressure on health and education services; unemployment; crime; air pollution and traffic congestion.*
2. *Example answer:* In Mumbai, Dharavi is the largest slum and is home to more than one million people, living in a very densely populated area. One of the urban planning strategies used by the government to improve the standard of living for those in the slum is a housing redevelopment project. This has involved the building of new houses. If residents are able to prove they have been living in Dharavi since 2000, the government has promised them a free house. This will enable many residents to have a house that is both larger and safer, providing people with more shelter and security. Alongside the construction of new homes, the government has also invested in improvements to water supplies and sanitation infrastructure, enabling the residents to access safe clean drinking water. This will contribute towards reducing waterborne illnesses and will improve hygiene in Mumbai.

Page 55 Distribution of UK population and cities

Quick quiz: 1. urban **2.** false
Questions:
1. (a) 0–600 people per sq km **(b)** *Any one of the following:* London; Manchester; Liverpool; Birmingham; Leeds.
(c) *One of the following:* they show abrupt changes across boundaries, which can be misleading; it can be difficult to distinguish between shades of colours.
2. The area of the highest population density in the UK is London, in the south-east of England. The north-west also has large areas with high population density, especially in and around Liverpool and Manchester. The remaining areas of high population density are in large cities, such as Cardiff in Wales, Birmingham in England and Edinburgh in Scotland.
3. Physical factors that can result in a higher population density include a temperate climate, and low-lying, flat, fertile land. Human factors that favour a higher population density include employment opportunities, and quality and availability of education opportunities. For example, cities with respected universities, such as Edinburgh and London, attract people to move there, contributing to a high population density. *Other answers might include: healthcare services; transport links.*

Page 56 Case study: Opportunities in the UK

Quick quiz: 1. integrated **2.** true
Questions:
1. *Example answer:* London has a huge variety of recreational opportunities, including world-class sporting events, a vast choice of shopping opportunities, museums, street performers, the West End theatre district, restaurants and bars. *Other answers might include: diverse cultural mix; employment opportunities; integrated transport systems.*
2. *Example answer:* An economic opportunity created by urban growth in London is the huge range of employment opportunities, which attract highly skilled workers from around the world. *Other answers might include: start-up companies; multinational companies located in the UK; very high property prices offer investment opportunities.*
3. *Example answer:* One of the positive impacts of migration on London is that migrants are contributing to the city's labour force. This benefits the city because it adds to the number of skilled workers, such as doctors and engineers. Between 2000 and 2011, European migrants made a net contribution of £20 billion to UK public finances. However, migration has also had negative social and economic impacts on London. A negative social impact is that in some areas there is racial tension between migrants and locals. A negative economic impact is that wealthy migrants have contributed to huge increases in house prices, with average house prices in Kensington and Knightsbridge exceeding £2 million. This is causing problems for local residents, because they are being priced out and may be unable to afford the inflated prices. *Other answers might include: population growth can cause housing problems; migration can increase cultural diversity.*

Page 57 Case study: Challenges in the UK

Quick quiz: 1. prices; commuter **2.** *One of the following:* loss of green spaces; farmland and gardens being built on; loss of wildlife habitat.
3. Expensive to clear the land
4. Cities may not have enough school places.
Questions:
1. *Example answer:* The Lower Lea Valley in London was selected for regeneration in order to reduce deprivation and improve living conditions.
2. *Example answer:* The urban wasteland of the Lower Lea Valley was cleaned and 9000 new homes were built. *Other answers might include: soil contaminated with industrial waste was washed and reused; accessibility was improved through construction of new land bridges; the Olympic Park created employment opportunities before and after construction.*
3. *Example answer:* One of the changes that has taken place in many UK cities has been the decline of manufacturing, and as a result the closure of many factories. This has led to many derelict factories, which are an eyesore and create dangerous areas. However, in recent years, many of these derelict factories have been converted into apartments for young professionals working in the cities. In recent years, this has been a challenge in London, where housing costs have increased, leading to increased inequality. For example, private rental prices have risen significantly over the last 10 years and there are huge differences in house price between different areas of London, leading to increased debt problems for many of the average earners in the city.
Other answers might include: above-average unemployment rates; commuter settlements; urban deprivation; dereliction; inconsistencies in the quality of healthcare.

Page 58 Urban sustainability

Quick quiz: 1. electric / hybrid car **2.** Singapore
3. *Any one of the following:* recycling, desalination, capturing and storing rainwater.
Questions:
1. Energy conservation is an important part of sustainable urban living. Solar panels can help to conserve energy by generating electricity and therefore reducing an urban area's reliance on the National Grid for its energy needs.
2. Planning green spaces in urban areas creates habitats for plants and wildlife, which may otherwise be unable to survive in urban areas. *Other answers might include: provide areas for outdoor recreation, are aesthetically pleasing and good for people's physical and mental health; provide areas for urban food production; contribute towards reduction of carbon emissions, because plants absorb carbon dioxide during photosynthesis.*
3. *Example answer:* Stockholm has introduced an electronic congestion charge scheme, which charges motorists for entering parts of the city on weekdays during peak times. This has resulted in reducing traffic levels by 22 per cent, contributing to decreasing carbon emissions. *Other answers might include: Hangzhou's public cycling system; Copenhagen's bike share scheme.*
4. *Example answer:* Sweden has established one of the world's best household waste recycling systems, with only 1 per cent of household waste ending up in landfill sites. Recycling stations within 300 metres of residential areas means that people can easily deposit their recyclable waste in special containers.

Page 59 Classifying development: economic measures

Quick quiz: 1. lower income country
2. newly emerging economy
Questions:
1. One economic measure of development is gross national income (GNI). GNI is the total income of a country. GNI can also be used to calculate GNI per capita by dividing the country's GNI by the size of the country's population. *Other answers might include: gross domestic product (GDP).*
2. Although economic measures of development are useful for comparing levels of wealth between countries, they do not show the inequality of wealth within a country. For example, a country with a high GNI per capita may have some very rich people, but a significant proportion of the population living in poverty. Also, economic measures do not adequately represent quality of life, which is important for determining a country's level of development. *Other answers might include: some countries might not have up-to-date data; the governments of some countries may not provide accurate data.*
3. (a) *For example:* Greenland; Indonesia; Papa New Guinea
(b) *For example:* Canada; Norway; Australia
(c) *For example:* UK; France; Japan

Page 60 Classifying development: social measures

Quick quiz: 1. per 1000 people **2.** children; first
3. life expectancy and number of years in education
Questions:
1. lack of access to healthcare
2. An advantage of social measures of development is that they give a reliable indication of the standard of living in a country. For instance, low death rates suggest that the healthcare system is very good. A limitation of social measures of development is that data provided by the governments of some countries may not be accurate, or there may not be accurate data available for an entire country in order to gain a measure, meaning not all measures can be used to compare all countries.
3. Very low, <0.42
4. The pattern is very uneven but, in general, the higher HDI figures are located in the south and far north of Africa, whereas the lowest HDI figures tend to be the landlocked countries in the middle of the continent.
5. One advantage of the HDI as a measure of development is that it combines economic and social measures, which together give a more comprehensive reflection of a country's development than using either measure separately.

Page 61 The Demographic Transition Model

Quick quiz: 1. Stage 1 **2.** Stage 4
3. Mexico – Stage 3; Indigenous rainforest tribes – Stage 1; the USA – Stage 4
Questions:
1. Declining birth rates suggest that a country has improved access to contraception and education, which leads to a reduction in the number of children born. *Other answers might include: a declining birth rate can suggest that a country has an established education system, which can result in more women choosing a career instead of having children; a declining birth rate suggests a country is becoming more developed, because fewer children are required to work.*
2. The total population starts to decline in Stage 5 of the DTM because young adults choose to have fewer children due to the costs associated with raising a child. *Other answers might include: the ageing population; desire for a large family decreases.*

3. Stage 2 of the DTM represents the early expanding stage, in which birth rates remain high and death rates fall dramatically, and therefore the overall population increases. One of the reasons for the fall in death rates is rapid improvements in access to and quality of healthcare in a country. An example is access to paediatric care, which affects the life expectancy of children, who are the most at-risk demographic group. Alongside improvements in healthcare, another reason for falling death rates is improvements in sanitation, reducing waterborne illnesses like cholera, and technological advances in food production, increasing access to food resources.

Page 62 Uneven development: causes and consequences

Quick quiz: 1. Figure 1 physical / historical
Figure 2 economic / physical **Figure 3** physical
Questions:
1. One historical cause of uneven development is colonialism. This has led to uneven development because resources from colonised countries are exploited for the benefit of the colonising nation. This prevented colonised countries from using their own resources to develop. Another historical cause of uneven development is conflict, because countries involved in lengthy conflicts are likely to have used large amounts of their resources on this, rather than investing in infrastructure, healthcare systems, or education. *Other answers might include: conflicts with high numbers of fatalities can reduce the number of people of working age.*
2. Extreme hazards such as earthquakes cause extensive damage to a country's infrastructure. For many LICs, these hazardous events inhibit their ability to develop due to the frequent devastation caused and the expense of rebuilding.
3. One impact of uneven development is international migration, as people in less-developed countries may leave their country of origin in search of a better quality of life, better employment opportunities, and better healthcare and sanitation, which contribute to a higher life expectancy. *Other answers might include: disparities in wealth and health within and between countries.*

Page 63 Reducing the global development gap

Quick quiz: 1. ...short-term aid. **2.** debt; invest
Questions:
1. Intermediate technology aims to meet the needs of local people by being simple to use and easy to maintain.
2. Investment from HICS or TNCs reduces the development gap in LICs by improving employment opportunities.
3. *Example answer:* More than 6.6 million tourists visited Brazil during 2016, which contributed $6.2 billion to the economy. This has enabled further investment in healthcare and education to improve basic services, and has provided more employment opportunities.
4. Fairtrade ensures that farmers in LICs are guaranteed a consistent price for the crops they produce. This enables farmers to improve their quality of life through a stable income.
5. Financial investment from TNCs or HICs can help LICs to improve infrastructure, making them better connected with better or more resources. This increases employment opportunities as people can travel and communicate more easily.

Page 64 Case study: Developing LICs and NEEs

Quick quiz: 1. (a) increase **(b)** reduce
(c) increase **(d)** increase
Questions:
1. *Example answer:* Trading policies have been changed in India, which has enabled the country to increase its volume of imports and exports and create strong trading partnerships with countries like China. This has contributed to improving India's economy.

2. *Example answer:* In India, TNCs have invested in telecommunications, contributing to the growth in the quaternary sector.
3. *Example answer:* India has a very large diaspora compared to other countries, with over 16 million people living abroad.
4. *Example answer:* India has two monsoon seasons that can adversely affect development. If the rainfall is too high, flooding can disrupt economic activities and infrastructure. India has also been ranked very low for its environmental quality, which could put off TNCs from investing in the country economically.

Page 65 Case study: Impacts of development

Quick quiz: 1. *One of the following:* more air pollution; more water pollution; traffic congestion; increased greenhouse gas emissions from rapid industrialisation.
2. *For example:* food and medical supplies.
Questions:
1. More people living and working in cities in NEEs has created more traffic congestion, which leads to longer travelling times.
2. *Example answer:* **Advantages:** Foreign investment has helped to improve infrastructure. It has provided more employment opportunities in India and also improved healthcare and education. *Other answers might include: the country has an improved position in the global market.*
Disadvantages: Foreign investment has led to poor working conditions and, particularly in cities like Mumbai, increased traffic and pollution.
3. *Example answer:* The rapid economic development of India in recent years has led to a variety of social, economic and environmental impacts, some positive and some negative. One of these is a difficulty providing enough housing for the growing population, with the government being unable to keep up with the demand. This has led to a rise in the number of squatter settlements in cities like Mumbai, with people using scrap material to build a form of shelter for their families. Another social and environmental impact of India's rapid development is rising levels of air and water pollution, which can lead to health problems. Another negative environmental impact of rapid development is traffic congestion. The roads in many cities have become gridlocked, with many commuters facing long hours stuck in vehicles when trying to navigate around the city. This has also contributed towards increased air pollution levels, which can lead to respiratory diseases.
However, a positive economic impact of India's economic development is that there are more job opportunities available. This has helped to reduce the number of people living in poverty. There is also better access to healthcare and education in some urban areas, which has helped to improve quality of life. In summary, rapid urban development in Mumbai has created many challenges, but as the economy grows, the quality of life for people living there is improving. *Other answers might include: changes to the aid the country receives and its impacts; increased demand for food; changing political and trading relationships with the rest of the world.*

Page 66 Causes of economic change

Quick quiz: 1. *One of the following:* farming; mining; fishing; logging.
2. Any service-industry job, such as teacher, sales assistant.
3. increased; countries
Questions:
1. One reason for deindustrialisation in the UK is globalisation, which has made it cheaper to import many products than to produce them in the UK. A second reason for deindustrialisation in the UK was mechanisation, which resulted in fewer manual workers being required as machinery was able to take their place.

Other answers might include: government policies, such as privatisation, resulting in a loss of jobs and a decline in traditional UK industry.

2. Deindustrialisation caused higher rates of unemployment in some areas of the UK, such as Wales and the north east, which previously had many jobs available in manufacturing. Meanwhile, the growth of the tertiary and quaternary sector have led to more jobs becoming available in the south of the UK, especially around London.

3. (a) 38%

(b)

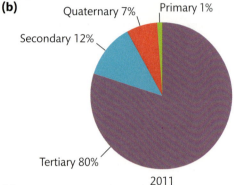

Quaternary 7% Primary 1%
Secondary 12%
Tertiary 80%
2011

(c) 34%

4. One reason the quaternary sector is growing is that the development of information technology, such as email and video conferences, has allowed UK companies to provide financial and business services to people around the world. *Other answers might include: more science / business parks have been established; the number of people doing research jobs has increased.*

Page 67 Impacts of industry

Quick quiz: 1. air pollution **2.** scarring of the landscape
Questions:
1. *Example answer:* Industries can operate more sustainably by limiting their environmental impacts. For example, Tata Steel conduct regular checks on water discharges to monitor the levels of suspended solids and hydrocarbons, both of which are pollutants. *Others answers might include: the Ultra-Low CO_2 Steelmaking (ULCOS) partnership; investment in reducing energy requirements and increased energy efficiency.*
2. *Example answer:* One of the ways industrial operations impact on the UK environment is increased concentrations of greenhouse gas emissions and local air pollution. For example, in 2013 Tata Steel in Port Talbot were issued with an enforcement notice because of black dust released into the atmosphere near residential areas close to the plant. Second, disposal of industrial waste in landfill can disrupt wildlife and pollute the soil, which can destroy wildlife habitats and can have wider impacts on the biodiversity of an area. Industrial units also spoil the beauty of the natural landscape, and the transportation of materials to and from industrial plants causes increased pollution from lorries.

Page 68 Rural landscape changes

Quick quiz: 1. *For example:* Northumberland
2. *For example:* the Outer Hebrides
Questions:
1. 68% **2.** approx. 2% (2–4% is acceptable)
3. Reason for growth: counter-urbanisation. *Other answers might include: many rural areas are attractive to retired people.*
Reason for decline: decline in the availability of services, such as GP practices. *Other answers might include: fewer jobs available in primary industries such as fishing.*
4. One social impact is rising numbers enrolled in local

services such as doctors' surgeries, which leads to longer waiting times. *Other answers might include: commuters increasing the volume of traffic, so longer travelling times for locals; pressure to build houses on greenbelt land leads to objections from local communities; second homes can have a negative effect on communities.*
5. One economic impact is a negative impact on the economy, which can lead to the closure of traditional local businesses. *Other answers might include: ageing population puts pressure on healthcare budgets; may be more healthcare jobs available, but they are often low paid.*

Page 69 Infrastructure improvements

Quick quiz: 1. population; transport
2. *For example:* Crossrail; HS2
Questions:
1. One benefit is increased employment opportunities and a second is more revenue for the UK economy. *Other answers might include: taxation*
2. *Example answer:* HS2 is an example of a railway improvement scheme. It intends to provide a high speed link between the north and the south of the UK to improve connectivity and reduce travelling time between regions.
3. (a) 200 million tonnes **(b)** 320 million tonnes
(c) Figure 1 shows that the overall imports have increased, which may have been because improvements to infrastructure have made the transport of goods easier and safer.

Page 70 The north–south divide

Quick quiz: 1. true **2.** economic and cultural
Questions:
1. approximately £185 per week
2. One of the reasons for the north–south divide in the UK is the decline of primary industry in the north and the growth of the tertiary and quaternary sectors in the south.
3. *One of the following:* life expectancy rates; average house prices; unemployment rates; public spending on infrastructure per resident.
4. *Example answer:* One strategy is the 'Northern Powerhouse' government strategy. This was developed to improve economic wealth in the north by investing in projects in cities such as Liverpool and Manchester and focused on improving sectors such as tourism. Another strategy is the proposed HS2 rail link, which will bridge the gap between the north and south by decreasing travel times and increasing the number of services available. *Other answers might include: Enterprise Zones, providing financial support; superfast broadband and straightforward set-up regulations to encourage new businesses in the north.*

Page 71 The UK in the wider world

Quick quiz: 1. voluntary; countries
2. *One of the following:* being able to trade within the single market; free movement of migrant workers.
3. true **4.** broadband; expand
Questions:
1. *Example answer:* The largest airport in the UK is London Heathrow, which serves 204 destinations, and was used by 78 million people in 2017.
2. Benefit 1: Increased tourism provides more employment opportunities. **Benefit 2:** Improved trading relationships boost the economy.
3. Trade between Commonwealth countries is on average 19 per cent cheaper than trade between member countries and non-member countries, because of similarities between legal systems and language.

4. (a)

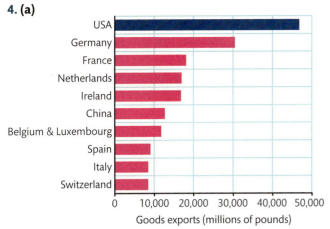

Goods exports (millions of pounds)

(b) France   **(c)** approximately £22 000 million

Page 72 Types of resources

Quick quiz: 1. Energy   **2.** oil
3. fresh water, food and energy
4. crop production and drinking
Questions:
1. (a) >2500– ≥3000 calories
(b) In wealthier countries people can afford to buy more food, and there tends to be a better infrastructure for distributing food. *Other answers might include: there is better access to transport, making it easier to buy food.*
2. Global inequalities in water supply can be the result of the amount of rainfall a place receives. For example, countries near the equator receive low amounts of rainfall every year, whereas places further north and south receive high amounts of rainfall.
3. One reason for global inequalities in energy supply is the different levels of wealth around the world. If a country is more developed, it is able to invest money in energy projects and import fossil fuels. *Other answers might include: physical factors, for example some countries have an abundance of geothermal energy resources; political factors, such as unrest in oil-producing areas; technology.*

Page 73 Food resources in the UK

Quick quiz: 1. *One of the following:* population growth; increased consumption.
2. ...increased.   **3.** more; prices
4. *For example:* avocados; fine beans; truffles.
Questions:
1. One reason for increasing food miles is the demand for high value imports from LICs, many of which cannot be grown in the UK's climate. *Other answers might include: rise in demand for organic produce; rise in demand for seasonal food all year round.*
2. Importing food products from countries further away increases food miles, creating larger carbon footprints. This is because the food is transported over a greater distance, which requires the combustion of fuel, releasing carbon dioxide.
3. C
4. *Example answer:* Agribusinesses involve running UK farms as businesses, cutting costs and maximising profits. This can help to meet the rising demand for food in the UK by allowing farms to maximise productivity and produce low-cost food whilst running a sustainable business. For example, Lynford House Farm, near Ely, is an example of an agribusiness. They share expensive machinery with other farms, and have secured external investment for the installation of a 54 million litre reservoir, making their food production more cost effective and sustainable.

Page 74 Water resources in the UK

Quick quiz: 1. rivers, reservoirs and groundwater aquifers

2. ...surplus.   **3.** transfer; surplus   **4.** person; increasing
Questions:
1. In the south and east of the UK, the demand for water is high due to a high population density, and lower rates of precipitation in the south east create serious water stress.
2. The Environment Agency constantly checks water quality for pollution and chemical levels, looking for things such as nitrates in the water, to ensure the quality is maintained for human consumption. *Other answers might include: the risk of pollution from industrial and agricultural areas; water pollution from run-off from roads.*
3. The map shows that the highest rainfall rates occur in the north-west of the UK, and the lowest rainfall rates tend to occur in East Anglia and the south-east. For example, western Scotland received 400–449 mm of rainfall, whereas East Anglia and the south-east only received 200–249 mm.
4. *One of the following:* western Scotland; north Wales and north-west England.
5. *One of the following:* northern Scotland; eastern Scotland; south-west England and south Wales.

Page 75 Energy resources in the UK

Quick quiz: 1. demand   **2.** It has decreased.
3. *Any two of the following:* solar power; wind power; biomass.
4. *One of the following:* cheap imports; concerns about CO_2 emissions from coal-fired power stations.
Questions:
1. Economic issues associated with exploiting energy sources include the cost of building nuclear power plants, which are very expensive to construct. Relying on nuclear power stations, rather than fossil fuels, to generate electricity is also more expensive. Environmental issues that may be caused by exploiting nuclear and fossil fuel energy sources include the problem of disposing of nuclear waste, which requires careful management, and the unsustainably high levels of carbon dioxide emissions released from fossil fuel-fired power stations, which contribute to global warming. There are also potential environmental issues associated with using renewable energy sources, such as flooding and habitat destruction which can be caused by building hydroelectric dams.
2. (a) UK wind power operating capacity has increased every year since 2000, increasing from 200 MW in 2000 to 13 500 MW in 2014–2015.
(b) 2000 MW   **(c)** 11 500 MW

Page 76 Distribution of food

Quick quiz: 1. insecurity
2. *One of the following:* rising population; economic development.
3. *One of the following:* they can afford to import lots of food; many HICs have high production levels; there is good distribution infrastructure.
4. calorie; LICs
Questions:
1. Climate variations can affect crop yields. Rising temperatures and extreme weather events can have a negative impact on crop yields. A second factor affecting food supply is poverty, as people living in poverty cannot afford fertilisers, equipment, or technology. *Other answers might include: technology; conflict; water scarcity; pests and diseases.*
2. *One of the following:* obesity problems; undernourishment.
3. (a) *One of the following:* the USA; China; India.
(b) D

Page 77 Impacts of food insecurity

Quick quiz: 1. population; security
2. true

Questions:

1. C

2. Food insecurity leads to subsistence farmers in many African countries using their land over and over again, in a process caused overcultivation. Lack of vegetation cover can expose the soils, which can be washed away easily. Eventually the land becomes infertile, which can cause soil erosion.

3. (a) When demand for food is high but supply is insecure, prices increase, especially for staple crops like rice.
(b) *One of the following:* rioting; conflict; increased world hunger.

Page 78 Increasing food supply

Quick quiz: 1. green; conditions
Questions:

1. *One of the following:* hydroponics; irrigation techniques; biotechnology.

2. *Example answer:* **Advantage 1:** The Indus Basin Irrigation System in Pakistan has helped to improve crop yields by giving farmers better access to water to irrigate their crops.
Advantage 2: The opportunities for fishing in the Mangla reservoir provide an additional source of food, which has improved the diet of local people.

3. *Example answer:* One disadvantage of the Indus Basin Irrigation System is that farmers can be deprived of water further downstream.

4. *Example answer:* One way food supplies can be increased is through the use of appropriate technology. For example, in drought-prone areas farmers use traditional water conservation techniques and planting methods. For instance, farmers using the Zai system in West Africa fill small holes with manure, which attracts termites that build tunnels that then capture water. The termites also recycle soil nutrients. This means that the farmers have been able to increase their water supply in a way that they can control. As a result, they can irrigate their crops and increase food supply without any damage to the environment. A second method of increasing food supplies is the use of drip irrigation techniques. Drip irrigation systems provide an efficient way to conserve water and can contribute towards yield increases of 30–50%. Also, low-cost microtube irrigation systems are easy to install and can be maintained by farmers. Both strategies lead to better growing conditions and therefore increased food supplies. *Other answers might include: biotechnology; the new green revolution; hydroponics.*

Page 79 Sustainable food options

Quick quiz: 1. true **2.** quotas **3.** permaculture
4. It is where people aim to reduce, reuse and recycle all of their waste.
Questions:

1. If more people buy locally produced, seasonal food, it reduces food miles, therefore lowering carbon emissions. It also supports local farmers.

2. *Example answer:* Urban farming initiatives can improve food supply sustainability by using recycled materials and new growing technologies to produce fresh fish, salads and herbs. For instance, GrowUp Box in London recycles old containers to grow fresh produce in an urban area. They can also reduce food miles by reducing the amount of food that needs to be brought into urban areas.

3. Permaculture involves simulating or directly using features of an ecosystem to improve crop yield. It is a sustainable food supply option as it involves working with the natural environment. This means that less pollution is released during the farming process.

4. Organic farming methods do not rely upon the use of chemicals, which can negatively impact soil health and pollinating insects, so it is more sustainable. This creates more environmentally friendly crops that have less impact on their surroundings.

Page 80 Distribution of water

Quick quiz: 1. security
2. *Any two of the following:* population growth; economic development; dietary changes increasing the demand for meat and dairy; increasing use of biofuels.
Questions:

1. Factor 1: Climate variations mean that some countries may have a water surplus, while others have a water deficit. For example, a hot climate with low annual rainfall will result in a water deficit. **Factor 2:** Poverty affects water availability, as LICs often do not have enough money to pay for mains water supplies. *Other answers might include: geology; pollution; over-abstraction; limited infrastructure.*

2. Over-abstraction of water from aquifers can lead to water insecurity because the natural water sources are unable to replenish quickly enough for reuse.

3. Most areas with a water deficit are located near the equator, because the climate is hot with low levels of precipitation. Most Middle Eastern and North African countries have a water deficit. However, there are also countries with a water deficit not located near the equator, such as Chile.

4. (a) 90 001–171 000 million litres per day
(b) Georgia, Oklahoma, or any other suitable state
(c) The climate of a state might affect the amount of water used. For instance, **Figure 1** shows that Texas and California, which have hot, dry climates, use the highest amount of water. *Other answers might include: the higher the population, the more water is used.*

Page 81 Impacts of water insecurity

Quick quiz: 1. true **2.** conservation; pollution
3. industrial; shortages **4.** false
Questions:

1. C

2. In some African and Asian countries, there is no reliable supply of clean fresh water. People are forced to drink contaminated water, which results in many people contracting waterborne diseases, such as cholera and dysentery.

3. Agriculture requires large amounts of water for livestock and irrigating crops. Water insecurity caused by drought can lead to problems with crop production, as there is not enough water to irrigate crops. This results in a reduction in crop yield.

Page 82 Increasing water supply

Quick quiz: 1. ...the removal of salt and minerals from a water source to produce fresh water.
2. water storage **3.** deficit
4. *One of the following:* flooding large areas of wildlife habitat; people forced to relocate; increased erosion downstream; blocking fish migration routes.
Questions:

1. *Example answer:* One advantage of the South–North Water Diversion Project in China is that it has the potential to increase agricultural and industrial supplies by providing a better water supply. Another advantage is that it provides a more reliable source of drinking water for the northern half of the country, improving the China's water security.

2. *Example answer:* One way water supplies can be increased is through the construction of dams, which help to control the river flow and create a large store of water known as a reservoir. The store of water can then be used for both domestic and industrial uses in local areas as required. Dams can be effective in maintaining a reliable supply of water. However, often large areas of land have to be flooded to make way for the reservoir, causing loss of habitat and displacement of settlements. Another way water supplies can be increased is by using the process of desalination. This is where special filters are used to remove the impurities from seawater to convert it to drinking water. It can greatly increase water supply in countries where the climate is hot with very low levels of rainfall. For example, the United Arab Emirates have invested in this process by building desalination plants to meet the demands of its residents in a barren landscape. Finally, water supplies can be increased by water transfer, where water is transferred from an area of surplus to an area of deficit through a network of pipelines. For example, the Kielder water transfer scheme in Northumberland transports water from the north to the south. This enables areas that have a water deficit to have a more reliable supply of water. Desalination and water transfer can be effective ways to maintain a reliable water supply. However, they are very expensive and are not always affordable for NEEs and LICs.

Page 83 Sustainable water options

Quick quiz: 1. installing a low-flush toilet; installing a smart meter
2. recycling **3.** sustainable
Questions:
1. Grey water reduces the total water use of households by using water from baths and showers to provide water for gardens.
2. A
3. *Example answer:* In Vimphere, Malawi, the NGO WaterAid installed a pump for the village which draws water from underground. An advantage of this scheme is that the villagers have access to safe water, which has significantly reduced the risk of them contracting waterborne diseases such as dysentery.
4. Water recycling is the reuse of treated water for a different purpose from its original use, reducing the quantity of water required. For example, a company may reuse water from one part of their production process in another, which contributes to reducing their overall water demands.

Page 84 Distribution of energy

Quick quiz: 1. supply; deficit **2.** 2 per cent
3. *One of the following:* physical factors such as climate, fossil fuel abundance, geothermal energy; the cost of exploitation and production; politics; technology.
4. highest
Questions:
1. Some countries have invested in technology. This has enabled them to extract new energy resources, like shale gas from fracking. A second reason for variations in energy supply is physical factors, such as climate and location. This is because some countries are more suited to the extraction of particular energy types, for example, geothermal energy in Iceland. *Other answers might include: cost of exploitation and production; political factors such as unrest in oil-rich areas.*
2. The global population is increasing and is expected to reach 9.87 billion people by 2050. Most of these people consume energy, increasing global consumption. In newly emerging economies, such as India and China, economic development increases industrial demand for energy, and as the country becomes wealthier the population has more money to buy electrical products and cars, all of which increase energy consumption.
3. (a) 25–49%
(b) *One of the following:* Niger; Chad; Democratic Republic of Congo. *Other answers might include: Mauritania, Guinea, Guinea-Bissau, Sierra Leone, Burkina Faso, Liberia, Central African Republic, Ethiopia, South Sudan, Kenya, Uganda, Rwanda, Burundi, Tanzania, Somalia, Madagascar, Malawi.*
(c) Figure 1 shows that mostly the countries with the worst access to electricity are the landlocked countries in central and east Africa, whereas countries with the best access are those that are on the coast in in the north and south of the continent.

Page 85 Impacts of energy insecurity

Quick quiz: 1. true
2. *One of the following:* wildlife habitats being damaged; disruption of indigenous people, fish and animals; oil spills that pollute water are difficult to clean up and kill wildlife.
Questions:
1. C
2. Uncertainty over the availability of fossil fuels in the future may reduce the availability of food products that require a lot of energy to produce. This will lead to considerable increases in food prices.
3. *Example answer:* Political unrest in the Middle East has resulted in uncertainties around oil supply and prices. Because energy is a key resource and supplies of fossil fuels are finite, it is likely there will be conflicts in the future about who controls fossil fuel supplies.
4. Energy is an essential resource for industrial production, so an inconsistent supply of energy can result in delays or closure of production lines. This has a negative impact on the economy.

Page 86 Increasing energy supply

Quick quiz: 1. false **2.** solar energy
3. It is non-renewable, because supplies of uranium are finite and will eventually run out.
4. *Any two of the following:* biomass, wind; hydro; tidal; geothermal; wave; solar.
Questions:
1. *Example answer:* **Advantage:** Oil extraction from the Athabasca tar sands in Alberta, Canada, provides employment for hundreds of thousands of people. This hugely boosts the local economy.
Disadvantage: Oil leaks have polluted the Athabasca River, which has damaged wildlife habitat.
2. *Example answer:* One way sustainable energy supplies can be increased is through the use of government incentives offered to people, so that it benefits people financially to install solar panels or small wind turbines to generate energy. Another way sustainable energy supplies can be increased is if companies and governments invest more money into renewable energy infrastructure, such as hydroelectric dams, wind farms and geothermal pipelines. This will increase global capacity to meet energy needs sustainably using renewable sources and reduce reliance on fossil fuels. A final way to increase sustainable energy supplies is investing in research into new renewable energy technologies, such as developing hydrogen vehicles, or researching the potential of algae to generate biofuels.

Page 87 Sustainable energy options

Quick quiz: 1. *One of the following:* cavity wall or loft insulation; using energy-efficient light bulbs and appliances; double or triple glazing; not leaving electrical products on standby.
2. *One of the following:* using public transport; walking, running or cycling instead of driving; car sharing; recycling; reducing waste; buying local produce.
3. consumption; further
Questions:
1. People can reduce their carbon footprint by using public transport, buying local produce, and recycling. These measures contribute towards reducing carbon emissions and decreasing the demand for energy, helping to manage energy sustainably.
2. *Example answer:* Hotel companies can adopt a 'no wash' policy for towels unless customers specifically ask for it during their stay. This means that the company will be washing towels less frequently and only when necessary, thereby using less

energy. *Other answers might include: companies providing recycling facilities and encouraging employees to use them; using video conferencing to reduce the need to travel to face-to-face meetings.*

3. *Example answer:* The Belo Monte Dam is an example of local renewable energy scheme in Brazil that is intended to provide the country with a sustainable supply of energy. Disadvantages of the dam include the fact that construction is expected to disrupt the lives of 20 000 people who live nearby, potentially displacing them from their homes. As well as disturbing the local tribes, the construction of the dam will also cause flooding in the Amazon rainforest which will disrupt the rainforest ecosystem and may harm or destroy animal and plant species. However, the Belo Monte Dam will also bring many advantages to Brazil. For example, the dam could generate enough electricity to provide power for approximately 20 million homes, helping to provide increased energy security. The construction of the dam will also provide employment opportunities for local people, reducing unemployment and boosting the local economy. On balance, despite the environmental problems, the Belo Monte Dam creates clean CO_2-free energy and will help to improve the economy and quality of life for many people in Brazil.

Page 88 Exam skills: Making geographical decisions

1. *Example answer:* Recovery from a natural hazard is almost always slower in LICs and NEEs than in HICs. This is due to a combination of factors, based on the immediate and long-term responses available to the country following the hazard event. Many wealthy countries have disaster plans in place in case of a hazardous event, enabling them to deploy help to those most in need faster than would be possible in an LIC that did not have plans in place. For example, after the earthquake in Ecuador in 2016 the country was able to initiate a state of emergency and deploy the army to help with the search and rescue. This resulted in a much lower death rate compared with the earthquake that occurred in Nepal, a less developed country, in 2015. More deaths occurred in Nepal because the country relied on emergency assistance from other countries, which took longer to be implemented. This demonstrates that level of development is the biggest factor in determining the speed of a country's recovery from a natural hazard. Furthermore, after the earthquake in Ecuador, the authorities constructed temporary shelters and mobile hospitals to provide warmth and food for the survivors following the event, which reduced the impact of the long term effects of the earthquake, such as the risk of diseases spreading and people losing their lives from injuries. Nepal did not have the resources to respond to the earthquake in the same way; they required help and support from funds raised by launching the Nepal Earthquake Flash Appeal, which took time to become available. These differing responses to the long term and secondary effects of the earthquakes are another reason why the death toll was much higher in Nepal than Ecuador. What is more, in more developed countries there is more money to invest in long term responses such as rebuilding homes and repairing major infrastructure such as roads and railway lines, and there may be government compensation schemes for the people affected that allow them to effectively rebuild their lives. All of these measures allow the country to recover recover from a natural hazard. In less developed countries, the government may be unable or unwilling to compensate people affected and pay for rebuilding, which can lead to people being homeless and travel infrastructure not being repaired for a long time after the hazard event, slowing the country's recovery. Overall, the evidence supports the view

that in most cases a country's recovery from a natural hazard is based largely on their level of development.

2. *Example answer:* Debt relief is the reduction or clearance of a poorer country's historical debt commitments to other countries, while fairtrade is an initiative that has established key principles to ensure farmers are consistently guaranteed a fair price for their produce. The debt relief initiative was launched in 1996 by the IMF and World Bank to assist poorer countries in helping to improve the lives of their citizens by removing some or all of their debt. Debt relief can be effective in reducing the development gap, as it allows LICs to spend more money investing in infrastructure, healthcare and education, all of which contribute to the country's level of development and benefit a large proportion of the population. However, the effectiveness of debt relief depends on the stability of a country's government – if the money released is not invested in key infrastructure to improve people's lives, it does not effectively reduce the development gap.

In comparison, although the fairtrade movement may initially seem entirely positive, it is not always beneficial for farmers, because the conditions set by the organisation can result in farmers having to pay considerable set-up costs to join the scheme. This can mean that the farmers are not benefitting from being part of the initiative in the first instance and so fairtrade is not doing anything to reduce the development gap. However, the scheme has brought about positive changes for many poor farmers with improved working conditions, tools and education to provide farmers with greater skills to maximise their crop yields, and giving them a secure income, which can contribute to reducing the development gap, although on a smaller, more localised level.

Page 89 Exam skills: Fieldwork

1. The environment is an an area with a safe accessible river, which is shallow enough to enter to take measurements of river depth, width and velocity. There are a footpath and footbridge to allow easy access to the river.

2. *Any one of the following:* river depth and width increases downstream; the load of the river becomes smaller and more rounded downstream; the velocity of the river increases downstream; the discharge of the river increases downstream.

3. *One of the following:* carry a phone for emergencies; carry warm and waterproof clothing; check tide times; be aware of weather; be aware of water depth, flow and currents; carry a first aid kit; be aware of unstable riverbanks.

4. *Answers might include:* limitations of your data collection methods; reliability of data; extent to which your conclusions are reliable; methods of improving your enquiry, such as different sampling methods, different data collection and recording methods.

Pages 90–94 Practice paper: Living with the physical environment

01.1 2000–3000 mm

01.2 The distribution of annual rainfall tends to be lowest in the south-east and highest in the north-west of the UK. **Figure 1** shows that parts of the south-east of the UK experienced the lowest average annual rainfall, with rates of less than 600 mm, whereas areas of north-west Scotland had an average annual rainfall rate of over 3000 mm.

01.3 Most of the upland areas of the UK are in the north and west, which leads to higher levels of annual rainfall, whereas the south-east is much flatter and closer to sea level, leading to lower levels of rainfall. *Other answers might include: prevailing winds and average temperature differences.*

01.4 One possible impact of the tropical storm is that buildings may be heavily damaged or destroyed, as shown in **Figure 2**, leading to people being made homeless. A second possible impact is that power lines may be destroyed, so people in the area may have no electricity for some time after the storm.

01.5

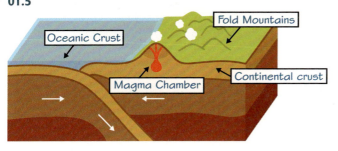

01.6 B

01.7 One reason why people choose to live in areas at risk from a tectonic hazard is that volcanic activity provides opportunities to harness geothermal energy. The minerals in volcanic ash create very fertile soil, increasing crop yields and therefore income for farmers. Another reason people live in tectonically active areas is tourism; tourists that come to see volcanic landscapes boost the local economy and provide job opportunities for local people.

In developed countries, evacuation plans mean that the risk to people is lower because people can be warned in advance of a hazardous event occurring. The infrequency of tectonic hazard events also means that some people believe that these events will not happen in their lifetime and so to them the benefits of remaining in the area outweigh the risks.

01.8 At constructive plate margins, convection currents in the mantle cause two tectonic plates to move away from each other. The separation of the plates causes a release of pressure and magma rises up through the Earth's crust. The magma wells up through the gap between the tectonic plates, forming volcanoes.

01.9 Monitoring and prediction techniques can hugely reduce the risks from a tectonic hazard. Monitoring techniques allow scientists to determine when a volcanic eruption is imminent. Thermal imaging equipment can detect the amount of heat coming from a volcano and chemical sensors can be used to assess gas levels in a volcano. GPS and tiltmeters monitor ground deformation, which can warn of an impending eruption. These monitoring techniques allow scientists to predict when a volcanic eruption is likely, normally in enough time to evacuate the area. This significantly reduces the risk of loss of life.

Seismometers can be used to record earthquake activity, but earthquakes are much more difficult to monitor and predict, although they tend to occur in predictable locations. Therefore, monitoring and prediction techniques can effectively reduce risks from volcanic hazards, but are not always effective in reducing the risk of earthquake hazards.

02.1 B

02.2 Tropical rainforests are hot all year round with average temperatures of around 28 °C. *Other answers might include: humidity; high levels of rainfall; very high biodiversity; most nutrients in decomposing matter; four layered vegetation.*

02.3 **Figure 4** shows that the largest areas of rainforest are in South America, Indonesia and West and Central Africa, and they are mostly located between the tropics. There are smaller areas of rainforest located in Australia.

02.4 **Figure 5** shows tropical rainforest that has been cleared to make space for a commercial palm oil plantation. Commercial plantations, such as this one, cause deforestation because they involve growing high-value crops in order to export them around the world. These include paper, rubber and especially palm oil, which requires very large areas of land and fertile soil. Rainforests are cleared by companies using a method called slash and burn, which quickly destroys huge areas of rainforest.

They then plant commercial crops, such as palm oil, on the cleared land and, because the soil quickly becomes infertile, they then destroy more rainforest to plant more crops. This is how commercial plantations lead to deforestation.

02.5

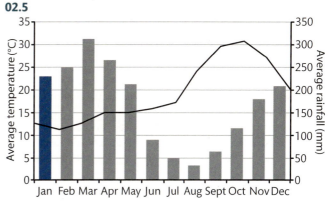

02.6 B

02.7 March

02.8 *For hot deserts, any two of the following:* tourism, such as camel trekking and 4×4 tours; extracting valuable minerals, such as copper; farming crops such as dates and fruit; generating solar energy.

For cold environments, any two of the following: extraction of oil; generating solar energy during the summer months; extracting valuable minerals such as gold and silver; commercial fishing; tourism, with attractions such as walking, sea cruises, skiing and fishing.

02.9 *For a hot desert, answers might include:*
- consideration of the extent to which a hot desert provides both opportunities and challenges.
- examples of opportunities: resource exploitation relating to agriculture, recreation and tourism.
- examples of challenges: remoteness; inhospitable conditions; environmental constraints.
- the relationship between the nature of the challenges and the desire / ability to overcome them in order for development opportunities to happen.

For a cold environment, answers might include:
- consideration of the extent to which a cold environment provides both opportunities and challenges
- examples of opportunities: mineral extraction; tourism; oil production; commercial fishing; generating solar energy and biomass energy.
- examples of challenges: extreme cold, as temperatures can reach −60 °C; remoteness and inaccessibility, as there is very limited transport and some areas are only accessible by reindeer or snowmobile; difficulties posed by the melting of the active layer in constructing buildings and infrastructure.
- the relationship between the nature of the challenges and the desire / ability to overcome them in order for development opportunities to happen.

03.1

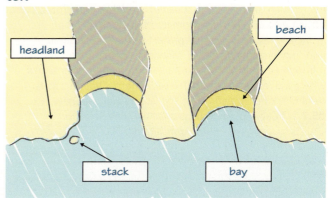

03.2 drift; prevailing wind; backwash

03.3 The formation of a spit, the landform shown in **Figure 9**, begins with prevailing winds causing the waves to push sediment up the beach at an angle. The backwash then brings material back down the beach at right angles to the coast. The swash and backwash process then results in longshore drift. When there is a change in the direction of the coastline, the transported material is deposited offshore. Over time the build-up of this deposited material off the coast causes a spit to form, stretching across from the headland. The spit curves at the end due to the effect of currents and secondary wind direction.

03.4 Hard engineering methods, such as the sea wall shown in **Figure 10**, can be used to protect the coastline from erosion by absorbing and reflecting wave energy. Sea walls are made of concrete, which is very strong and durable. They are often a curved shape, which helps to reflect wave energy away from the coastline. Rock armour and gabions use large boulders and wired cages of rocks to absorb wave energy, which prevents the waves eroding the land behind them. Groynes are a hard engineering method that protects the coast against the effects of longshore drift by preventing sand from being washed away.

04.1 vertical; banks; hydraulic

04.2

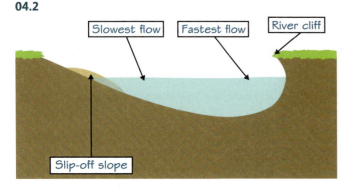

04.3 The landform shown in **Figure 12** is a meander. The flow of the water swings from side to side, creating a corkscrew motion known as helicoidal flows. The current moves quickly in the outside of the bend and the force of this current against the bed and the banks begins to wear them away, a process known as hydraulic action. Additionally, the material being carried in the flow scrapes against the bed and banks in a process called abrasion which increases the rate of erosion. The bank of the river becomes undercut, collapses, and a small river cliff is formed. On the opposite side of the river, sand and shingle are deposited due to the slow current and high levels of friction. This means that there is not enough energy to continue to transport the material and so it is deposited. Over time this leads to the formation of a slip off slope, which is a build-up of deposited material.

04.4 Hard engineering can protect settlements from the effects of river flooding in several ways. Dams and reservoirs, like the one shown in **Figure 13**, restrict the flow of a river, which decreases the speed of the water and protects settlements downstream from flooding. Embankments are artificially raised banks that protect settlements from river flooding by allowing a river channel to hold more water before it floods. For this reason, they are often built on river banks where a river passes through a settlement. *Other answers might include: straightening and flood relief channels.*

05.1 retreat; slip; material

05.2

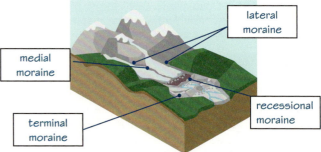

05.3 **Figure 15** shows a tarn lake inside a corrie. A tarn lake begins with the formation of a corrie, starting with the accumulation of snow in a hollow. The size of the hollow increases as the snow becomes compacted under the weight of further snowfall. Processes such as freeze-thaw weathering, where the action of water freezing causes it to expand and exert pressure on the rock, lead to cracks. Over time, plucking causes the back wall of the corrie to steepen and rotational slip and abrasion cause the hollow to deepen. At the front of the corrie, where erosion is at its lowest, materials are deposited and moraine builds up to form a raised lip. When the ice inside the corrie melts, it leaves a lake known as a tarn.

05.4 Tourism can have both positive and negative impacts on glaciated landscapes. People come to glaciated landscapes to relax, see the spectacular views and participate in outdoor activities such as hiking, climbing and mountain biking. This boosts the economy of the area, as it brings more customers to local businesses and provides jobs for local people working as tour guides and in outdoor activity centres. However, these jobs are often seasonal, which is a negative economic impact of tourism. Another negative economic impact is that the demands for holiday rental properties can increase house prices, making it difficult for local people to afford to live there. Tourism can also have a negative environmental impact on glaciated upland landscapes, contributing to footpath erosion and littering. *Other answers might include: tourists may disturb fragile wildlife habitat; tourism can lead to pressure to develop an area; tourists can contribute to traffic congestion, increasing commuter time; tourism can encourage conservation efforts.*

Pages 95–99 Practice paper: Challenges in the human environment

01.1 46%

01.2 54% – 39% = 15%

01.3 One factor that causes rapid growth of urban areas is rural–urban migration, where people move from the rural areas to urban areas, often because urban areas tend to have more employment opportunities and better access to healthcare and education services.

01.4 8397

01.5 D

01.6 *One of the following:* Employment opportunities; places with more and better paid jobs tend to have higher population density; quality and availability of healthcare; quality and availability of education services, such as schools and universities; proximity to major roads and airports; leisure opportunities such as shops, restaurants or sports facilities.

01.7 *Example answer:* One challenge created by urban growth is urban sprawl, where cities expand outwards into the surrounding countryside. This increases the demand for land on the rural–urban fringe; developers want to construct new housing developments, leading to green spaces, such as agricultural land and large gardens, being sold off for development. It also increases house prices on the rural–urban fringe, which means people may have to move further to find affordable housing and commute long distances. Urban growth can also create challenges relating to housing and inequality within the city. For example, in London, although many properties are worth millions, it is estimated that 27 per cent of Londoners live in poverty, partly because housing costs in the city are so high. **Figure 2**, the map of Manchester, also shows a complete lack of green space. This is a challenge which is associated with urban growth, as it can have a negative impact on the environment and people's quality of life. *Other answers might include: lack of school places; inconsistencies in healthcare; waste disposal problems; traffic congestion; increase in commuter settlements; job opportunities.*

01.8 It is located on a brownfield site, which is a sustainable use of land, providing more homes without encroaching onto green belt around the edge of the city. *Other answers might include: the proximity to public transport links.*

01.9 *Example answer:* **Figure 3** shows a sustainable urban development in London. This development shows several strategies that are necessary for sustainable management of resources and transport. The development is near to public transport, reducing the need for car ownership. Cars produce greenhouse gases and cause traffic congestion, so it is necessary for sustainable transport management to offer alternatives to cars. Other cities, such as Hangzhou in China, have popular bike sharing schemes to reduce the number of people using cars. Strategies used to sustainably manage resources include conserving water by recycling grey water for energy production, catching and storing rainwater, and reclaiming and purifying water from sewers. *Other answers might include: strategies to conserve energy (solar panels, insulation in new buildings to reduce demand); waste recycling; creating green space.*

01.10 *Answers should include:*
- the effectiveness of a specific urban regeneration project and how successful it has been in helping to attract people to an area.
- urban regeneration strategies: cleaning polluted soil, building new homes, developing brownfield sites, creating jobs, and increasing accessibility.
- a judgement of how successful the project has been.

02.1

02.2 $16 230–29 290

02.3 GNI per capita doesn't show the inequality of wealth within a country. A country with a high GNI per capita may have a few very rich people and a significant proportion of the population living in poverty.

02.4 One cause of uneven development is the geographical position of a country. If a country is landlocked, this limits their ability to trade, affecting their economic growth. Another reason for uneven development is access to physical resources. Some countries have valuable natural resources. For example,

some Middle Eastern countries have a large supply of oil, which is valuable for trading with other countries. This enables their economies to develop. *Other answers might include: extreme weather; conflict; trade; debt; colonialism.*

02.5 approximately 68%

02.6 approximately 3%

02.7 One of the main reasons for the UK's changing employment structure, as shown in **Figure 5**, is deindustrialisation. The proportion of the work force employed in the secondary sector, manufacturing, has steadily declined between 1961 and 2011. This is because machinery and computers have replaced people and because globalisation means it is now cheaper to import many products than to produce them in the UK. Another change in the UK's employment structure is an increase in the tertiary sector. Reasons for this include advances in technology, such as email and video conferencing allowing UK businesses to provide financial services to businesses around the world.

02.8 One strategy used to solve the regional differences between the UK is the Northern Powerhouse initiative launched in 2016. The government is investing and supporting businesses in cities such as Liverpool and Manchester, to help expand their economies and increase the number and variety of jobs available. Another strategy is the planned HS2 rail development, which will provide a high-speed link between the north and the south of the UK. This will reduce the commuting time between key cities in the south and the north.

02.9 Rapid economic growth can cause difficulties providing enough housing for the growing population in urban areas, which can lead to increasing land and house prices and more squatter settlements. *Other answers might include: growth in development, such as construction.*

02.10 *Answers should include:*
- comparison of the effectiveness of several strategies and how successful they have been in reducing the development gap.
- a range of strategies: fairtrade, microfinance loans, debt relief, industrial development, aid, tourism, investment, and intermediate technology.
- reference to an example of how a strategy has helped to reduce the development gap in one place.

03.1 A

03.2 **Figure 7** shows that the UK's reliance on solid fuels, such as coal, for primary energy consumption has decreased since 1970, whilst the proportion of primary energy consumption made up of electricity, including from renewables, has gradually increased from only 2% in 1970 to over 10% of total primary energy consumption in 2014.

03.3 There have been concerns about carbon emissions from fossil fuels, which has led to an increase in using renewable energy sources and renewable energy technology has improved, making renewable resources an increasingly viable source of energy.

03.4 Drilling for oil and gas in remote and environmentally sensitive areas such as the Arctic can disrupt wildlife habitats and may lead to oil spills, which are difficult to clean up and can pollute water and harm indigenous wildlife. *Other answers might include: birds killed by wind turbines; habitat destroyed by hydroelectric flooding; greenhouse gases emitted by burning fossil fuels; decrease in biodiversity caused by commercial biofuel plantations.*

03.5 Although fossil fuels currently provide most of the UK's energy, the UK ought to considerably decrease its reliance on fossil fuels and increase the amount of energy generated from renewable sources. The UK is dependent on importing fossil fuels from other countries, and whilst this is currently a cheap way to supply energy, fossil fuels are finite, so as supply decreases the price will inevitably rise. Also, burning fossil fuels is one of the biggest causes of climate change: it is unsustainable and bad for the planet. Renewable sources of energy, such as solar and wind energy, do not produce greenhouse gases and are not a finite resource. Therefore, investing in them would secure the UK's energy supply for the future.

04.1 B

04.2 The map shows that the majority of countries with a high percentage of undernourishment are in Africa. For example, in nearly all countries on the east coast of Africa 25% or more of the population are undernourished.

04.3 Rising food prices can decrease food security by making it more difficult for people to afford sufficient nutritious food.

04.4 *Example answer:* The Indus Basin Irrigation System in Pakistan, which includes over 85 dams and 12 canals, is a large-scale example of a strategy used to increase food supply. Advantages of the scheme include the fact that it has improved yields of a variety of crops and increased the range of crops that can be grown. However, disadvantages of the scheme include farmers further downstream being deprived of water. It has also caused the soil to become waterlogged in places, which limits what can be grown in these areas. Another strategy to increase food supply is aeroponics, in which plants are grown without soil and sprayed with water and nutrients. This has increased yields and reduced costs for farmers. *Other answers might include: the new 'green' revolution; biotechnology; drip irrigation; hydroponics; appropriate technology.*

05.1 C

05.2 The map shows that countries with high water stress tend to be in the Middle East and north Africa. For example, the United Arab Emirates has an extremely high water stress of over 80%.

05.3 Conflict can lead to an inconsistent supply of water, particularly when dams in one country affect the water supply in another country. An inconsistent supply creates water insecurity.

05.4 *Example answer:* The South–North Water Diversion Project in China, the biggest inter-basin water transfer project in the world, is a large-scale example of a strategy to increase water supply. It provides a more reliable source of water for the north of China, increasing supplies available for agriculture and industry. However, disadvantages of the scheme include the disruption to wildlife and the relocation of around 330 000 people. Another strategy to increase water supply that has advantages and disadvantages is desalination. Its advantages include that it can provide water for very arid areas, such as the Middle East, but its disadvantages are that it is very expensive and it contributes towards carbon emissions. *Other answers might include: dams and reservoirs, diverting supplies and increasing storage.*

06.1 A

06.2 Countries with the lowest percentage of electricity generation from renewable sources tend to be located on the eastern side of the world. For example, Asia has a large percentage of countries with a 0–19.9% of electricity generation from renewable sources, whereas much of South America generates over 60% of its electricity from renewables.

06.3 Conflict can cause energy insecurity because many countries rely on importing fossil fuels to meet energy demands and unrest may disrupt supply. For example, unrest in the Middle East, a key provider of oil, has led to uncertainties around global supply and prices.

06.4 *Example answer:* Strategies used to increase energy security have both advantages and disadvantages. The main advantage of investments in renewable sources of energy, such as hydroelectric power and solar, is that they are not finite resources, so they will never run out. This helps to guarantee a secure supply of energy for the future. Another main advantage of using renewable energy to increase energy security is that they produce less greenhouse gases than fossil fuels, making them a more sustainable source of energy. Nuclear can be an effective strategy of increasing energy security because, unlike fossil fuels, supplies of uranium will not run out in the foreseeable future and it is a relatively cheap source of energy. However, a disadvantage of this strategy is that disposing of nuclear waste can have negative environmental impacts. *Other answers might include: fossil fuels are relatively cheap; burning fossil fuels produces greenhouse gases; negative environmental impacts of renewable energy sources such as HEP flooding.*

Pages 102–103 Practice paper: Geographical applications

01.1 The average temperature of the planet has risen steadily since 1860. In 1860, it was less than 13.6 °C, but by 1990 it had increased to around 14.3 °C.

01.2 Car ownership around the world has steadily increased, which has increased the amount of carbon dioxide from transport released into the atmosphere.
Other answers might include: population increase; deforestation; agriculture.

01.3 **Figure 1** shows a clear correlation between the amount of carbon dioxide in the atmosphere and the average global temperature; both have increased steadily since 1860. Most scientists agree that an increase in global temperatures at this rate is not sustainable. It may result in water and food insecurity, due to climates around the world changing and affecting the amount of water available and what crops can be grown. It is also likely to cause increasingly frequent and destructive extreme weather events, contributing to widespread loss of habitat and mass extinctions. Therefore, reducing carbon emissions, and thereby limiting the effects of climate change, is essential to the long term sustainability of our planet.

01.4 *One of the following:* more frequent and more destructive extreme weather events; extinctions, as climate and weather changes and species are unable to adapt quickly enough; warming oceans, threatening fish stocks and causing coastal flooding.

02.1 B

02.2 Climate change is having a wide variety of impacts on both the environment and people. As **Figure 2** shows, rising temperatures are making it difficult for many animal species to find enough food to survive. Melting ice has meant that polar bears have less time to find food, resulting in fewer polar bear cubs surviving into adulthood. If climate changes continue, leading to even faster loss of Arctic ice coverage, polar bears may become extinct. Climate change is also affecting people. Rising sea levels have increased the risk of coastal flooding in many areas, causing loss of life, destroying buildings, and limiting economic developments.

As **Figure 2** shows, the Maldives are very low lying islands, and the negative impact of climate change on both the fishing and tourism industries there is making life on the islands more difficult. Ultimately, if climate change continues and sea levels keep rising, the islands may be completely submerged. *Other answers might include: extreme weather events; food insecurity; water insecurity; habitat loss.*

03.1 Electric cars, shown in **Figure 4**, do not emit carbon dioxide or other greenhouse gases, so if many people used them the rate of climate change would slow down by reducing the amount of carbon dioxide emitted.

03.2 Carbon capture and storage pose potential risks to human health and the environment because of the possibility of CO_2 not being contained. There are concerns about leakage during the capture process and once the CO_2 has been stored it could leak into groundwater supplies. This would increase the acidity of the water, making it dangerous for human consumption and harming any river ecosystems that might become polluted.

03.3 *Answers should include:*
- evaluation of the strategies in relation to the challenges caused by responding to climate change.
- clear references to the particular effectiveness of carbon capture and storage as climate change responses, rather than generic observations about reducing carbon emissions.
- supportive evidence, including from the source in the resource booklet, and detailed links between content from different areas of the course of study.
- clear argument either for or against the efficacy of carbon capture and electric vehicles.

Answers might include that it is evident that many different responses are required to effectively respond to climate change, consequently any particular strategy is only likely to tackle some of the causes of climate change.

04.1 *Any two of the following:* traffic counts, pedestrian counts, environmental quality surveys, land use mapping, building heights, photographs, field sketches.

04.2

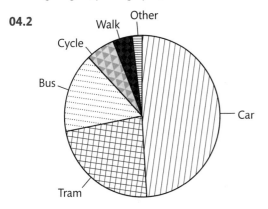

04.3 pictograms and bar charts

04.4

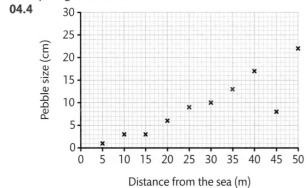

Distance from the sea (m)

04.5 Pebble measurements may not be accurate because the students may have measured the wrong part of the pebble or misread the value. *Other answers might include: distance from sea may not have been a straight line, so inaccurate; students may have recorded measurement data inaccurately.*

04.6 pebble size total = 92, divided by 10 = 9.2 cm

04.7 Figure 8 shows a correlation between change in pebble size and distance from the sea; in general, the greater the distance from the sea, the larger the pebble size. For example, the graph shows that at 5 metres from the sea, the pebble sample was 1 cm, but at 50 metres from the sea, the pebble sample was 22 cm. The exception to this is the anomalous result in sample 9. The graph shows that this is an outlier which does not fit with the trend on the rest of the graph.

04.8 *Any two of the following:* carry a first aid kit; wear appropriate footwear; be aware of tide times or areas of deep water; carry warm and waterproof clothing; carry mobile phones for emergencies; do not approach livestock.

05.1 *Student's own answer. Answers might include:*
- safety considerations: low water level, slow water flow
- range of survey points available
- range of survey points available with enough variation to show changes over distance.

05.2 *Student's own answer. Answers might include:*
- linking method to concept / theory, such as river channel depth linked to changes in a river's long and cross profile
- explanation of sampling strategy linked to quality of data.

05.3 *Student's own answer. A detailed answer will demonstrate a balanced appreciation of the effectiveness of components in the named presentation technique. Answers might include:*
- data presented and type of presentation used
- which variables are involved and how they are represented, for example, with units, scales and values on axes.

05.4 *Student's own answer. Answers might include:* specific reference to the methods, results and conclusions to provide a detailed evaluation as to how the geographical enquiry can be improved; evaluation of all three elements of the geographical enquiry; measuring and recording data; processing and presenting data, and evaluation and conclusion.

Published by BBC Active, an imprint of Educational Publishers LLP, part of the Pearson Education Group, 80 Strand, London, WC2R 0RL.

www.pearsonschools.co.uk/BBCBitesize
© Educational Publishers LLP 2019
BBC logo © BBC 1996. BBC and BBC Active are trademarks of the British Broadcasting Corporation.

Typeset and illustrated by Newgen KnowledgeWorks Pvt. Ltd., Chennai, India
Produced by Newgen Publishing UK
Cover design by Andrew Magee & Pearson Education Limited 2019
Cover illustration by Darren Lingard / Oxford Designers & Illustrators

The right of Michael Chiles to be identified as author of this work has been asserted by him in accordance with the Copyright, Designs and Patents Act 1988.

First published 2019

22 21 20 19
10 9 8 7 6 5 4 3 2 1

British Library Cataloguing in Publication Data
A catalogue record for this book is available from the British Library.

ISBN 978 1 406 68596 1

Printed and bound in Slovakia by Neografia.

The Publisher's policy is to use paper manufactured from sustainable forests.

Text Permission:
Pearson acknowledges use of the following extracts:
p13: Environment Agency : The Environment Agency; 2013. http://apps.environment-agency.gov.uk/wiyby/37837.aspx Contains public sector information licensed under the Open Government Licence v3.0; **p13: The Met Office :** The Met Office; 2015, https://www.mirror.co.uk/news/uk-news/uk-heatwave-here-blowtorch-heat-5970341. Contains public sector information licensed under the Open Government Licence v1.0; **p15: NASA:** Satellite sea level observations; NASA Goddard Space Flight Center, https://climate.nasa.gov/embed/128/; **p21: World Resource Institute** Indonesia's Tree Cover Loss Slows Substantially after Previous Highs; World Resources Institute; 2015. Creative Commons Attribution 4.0 International License; **p52: United Nations:** United Nations, Department of Economic and Social Affairs, Population Division(2015). World Urbanization Prospects: The 2014 Revision, (ST/ESA/SER.A/366).**p55:Office for National Statistics:** Office for National Statistics. Contains public sector information licensed under the Open Government Licence v3.0; **p59, 96, 123: World Bank:** GNI per capita; Atlas method (current US$); The World Bank Group. Available under Creative Commons Attribution 4.0 International license (CC-BY 4.0); **p68: Office for National Statistics:** 2011 Census Analysis - Comparing Rural and Urban Areas of England and Wales; Office for National Statistics. Contains public sector information licensed under the Open Government Licence v3.0; **p69: Office for National Statistics:** Who does the UK trade with. Office for National Statistics. https://www.ons.gov.uk/businessindustryandtrade/internationaltrade/articles/whodoestheuktradewith/2017-02-21. Contains public sector information licensed under the Open Government Licence v3.0; **p71: Office for National Statistics:** Office for National Statistics. Contains public sector information licensed under the Open Government Licence v3.0; **p70:**

Office for National Statistics: Annual Survey of Hours and Earnings: 2017 provisional and 2016 revised results; Office for National Statistics. Contains public sector information licensed under the Open Government Licence v3.0; **p74: The Met Office :** UK Climate; https://www.metoffice.gov.uk/public/weather/climate; Met Office. Contains public sector information licensed under the Open Government Licence v1.0; **p75: Office for National Statistics:** Digest of UK Energy Statistics (DUKES): renewable sources of energy; https://www.gov.uk/government/statistics/renewable-sources-of-energy-chapter-6-digest-of-united-kingdom-energy-statistics-dukes#history; Office for National Statistics. Contains public sector information licensed under the Open Government Licence v3.0; **p76:World Bank:** Data Bank: World Development Indicators; http://databank.worldbank.org/data/reports.aspx?source=2&series=AG.PRD.FOOD.XD&country=#,%20which%20I%20believe%20is%20CC4.0,%20though%20please%20confirm; The World Bank Group. **p80: U.S. Geological survey:** Dieter, C.A., Maupin, M.A., Caldwell, R.R., Harris, M.A., Ivahnenko, T.I., Lovelace, J.K., Barber, N.L., and Linsey, K.S.; Estimated use of water in the United States in 2015: U.S. Geological Survey Circular 1441, 65 p., https://doi.org/10.3133/cir1441 found in Water Use in the United States, https://water.usgs.gov/watuse/wuto.html; US Geological Survey; 2018; **p84: OECD/IEA:** Africa Energy Outlook, International Energy Agency (c), https://www.iea.org/publications/freepublications/publication/WEO2014_AfricaEnergyOutlook.pdf; © 2014. Used with permission of IEA publishing **p95: United Nations:** United Nations, Department of Economic and Social Affairs, Population Division, World Urbanization Prospects: The 2014 Revision, Highlights (ST/ESA/SER.A/352), © 2014, United Nations; **p96: World Bank:** GNI per capita, Atlas method (current US$). The World Bank Group. Available under Creative Commons Attribution 4.0 International license (CC-BY 4.0); **p99: US Energy information Administration:** International Energy Statistics, Eia.gov 4 March 2018; US Energy Information Administration; **p100: The Met Office:** Global Surface Temperatute, https://www.metoffice.gov.uk/research/monitoring/climate/surface-temperature; Met Office. Contains public sector information licensed under the Open Government Licence v3.0; **p101: Intergovernmental Panel on Climate Change:** IPCC, 2014: Summary for Policymakers. In: Climate Change 2014: Mitigation of Climate Change. Contribution of Working Group III to the Fifth Assessment Report of the Intergovernmental Panel on Climate Change [Edenhofer, O., R. Pichs-Madruga, Y. Sokona, E. Farahani, S. Kadner, K. Seyboth, A. Adler, I. Baum, S. Brunner, P. Eickemeier, B. Kriemann, J. Savolainen, S. Schlömer, C. von Stechow, T. Zwickel and J.C. Minx (eds.)]. Cambridge University Press, Cambridge, United Kingdom and New York, NY, USA.

Photographs
Shutterstock: austinding 1, Multiverse 11,90, Felix Mizioznikov 12, Steve Allen 14, Jacob Lund 22, liveyourlife 26, Tatiana Popova 27cl, Peter Turner Photography 35, Ann in the uk 39cr, stocker1970 50, jamesdavidphoto 51cr, CRS PHOTO 53, Damian Pankowiec 54cr, Pete Holyoak 89, Rich Carey 91, John Lumb 92br, Flyby Photography 93cl, Yavuz Sariyildiz 97, FloridaStock 100, Ekkapan Poddamrong 101tl, VVO 101cl, Scharfsinn 102, **Alamy Stock Photo:** Jake Lyell 25, Alexander Lutsenko 27c, Chris Cole 33, Lightworks Media 34, Robert Morris 39br, AlanWrigley 43, Marc Hill 45, Dave Ellison 48, Andrea Obzerova 49, Washington Imaging 51tr, WorldTravel 54tr, Anton Gvozdikov 58tr, david gregs 58cr, Evren Kalinbacak 62tc, Imagebroker/ Gerhard Zwerger-Schoner 62tr, robertharding/Gavin Hellier 65, Pearl Bucknall 67tr, A.P.S. (UK) 67tl,92bl, Pat Canova 78, Atstockfoto 93cr, Jeff Tucker 94cl, The Photolibrary Wales/Kevin Richardson 94cr, Stan Kujawa 96, Steve Tucker 103, **Crown copyright and database rights (2018) OS 40130946:** 29,36,47,95

Note from the publisher
Pearson has robust editorial processes, including answer and fact checks, to ensure the accuracy of the content in this publication, and every effort is made to ensure this publication is free of errors. We are, however, only human, and occasionally errors do occur. Pearson is not liable for any misunderstandings that arise as a result of errors in this publication, but it is our priority to ensure that the content is accurate. If you spot an error, please do contact us at resourcescorrections@pearson.com so we can make sure it is corrected.

Websites
Pearson Education Limited is not responsible for the content of third-party websites.